Answering the Call

Answering the Call

◆

Communication Technologies and the Not-for-Profit Organization

Joseph R. Liberto

iUniverse, Inc.

New York Lincoln Shanghai

Answering the Call
Communication Technologies and the Not-for-Profit Organization

iUniverse, Inc.

For information address:
iUniverse, Inc.
2021 Pine Lake Road, Suite 100
Lincoln, NE 68512
www.iuniverse.com

ISBN: 0-595-32385-5

Printed in the United States of America

This book is dedicated
To those who need,
and to those who care.

Especially to the hardworking folks of UCAN,
The Abraham's Tent Coalition,
and all of the likeminded organizations
for whom this book is intended.

This book is dedicated also to my friend Paul Lohinski,
a man who loves the telephone much more than I do.

With Special Thanks and Love
To Mandy, who put up with the late-night typing,
and to our wonderful parents and families.

and finally,
To Chuck "7 and 7" Clarke, Bob Hettchen,
Bernie, Dan, Ed,
"Patrick the Seminarian",
Karen, Annette, and Maureen
and all my wonderful and unique friends and family of St. Joseph Parish
who have truly made this work possible, and who have proven that
the journey is *the reward.*

Contents

Section 1: The Changing Role Of Telecommunications In Not-For-Profit Organizations

Section 2: Features, Features, Features

Section 3: Installation, And Everything After

Today, a friend of mine whose name is Ed, and myself felled a large Oak tree. We really didn't intend to cut down the Oak. We endeavored to put the 50 year-old pine—which had suffered some damage due to an unfortunate round with lightning—out of its prolonged misery. The pine tree lived between the Oak tree and the ravine of a dry stream bed on St. Joseph's backyard. The Oak happened to be in the way.

Like many things in life, taking down a large tree means that you can never be quite certain of which way things are going to go. This was, unfortunately, the untimely fate of the grand oak tree. It wasn't that we didn't plan, or prepare, or know how to fell a tree. We both have done this before, and had the proper riggings, saws, and preparations. The chains were checked, and the blades were sharpened. We looked at the weather, the wind, the slant of the tree. We cabled the pine tree, and sawed, and pulled. We sawed some more, and pulled some more. The mighty pine wouldn't budge. The sound of chainsaws could be heard for half a mile. Cracks and pops could be heard, but to no avail, the mammoth tree remained steadfast.

Finally, when there seemed that nothing was left of the foundation of the old tree's trunk, it began to move. It was supposed to go dead left, into a ravine. It went slightly left, then slightly right, then wildly, right into the oak. Well, so much for planning.

Organizations small and large are like this, I think. One plans, leads, and makes the decisions which one, as a leader, feels to be prudent and informed. Small changes occur in due course, some are picked up by the "sensors" of the management, others go undetected. It is but a fallacy to think that we can predict the world. The patterns of life are indeed grand, much more so than the years or lives of individual men. Even with the leveraging hand of technology, thinking that the world, with all of its variation both natural, economic, and political, can conform to routinization or predictability is ludicrous as thinking that two 250 pounds of manpower can drop a three-ton tree at a 90° angle every time. Just ask the old Oak tree, because I don't think it saw that one coming either.

A sad but illustrative story this is, and how appropriate for today's non-profit. What does this have to do with technology?

I think about this a lot when I think about the future. One day, I want to start my own organization, and use my technical background to help other non-profits leverage technology for the purposes of improved communication. An organization which helps other organizations bring to market their philosophy, and ideals. In this fashion I endeavor to help others work more efficiently and more intelligently.

Nothing is more intriguing to me than using ingenuity to solve some kind of insurmountable problem, like those geared-down pulleys that riggers use to move tons and tons of equipment and metal at shipyards. The leverage of Information Technology is the chain pull of the age and era in which we live. Just as an individual rigger can, using gears and hydrolics, move 5 tons of metal over yards and yards of ship, or move a large tree, an organization can conveive an impossible yet routine activity, like for instance a mass mailing, and using a simple tool like a database can easily make it happen. Without the knowledge of implementing this simple tool, most non-profits would not be able to pull this kind of thing off.

One basic leverage device that virtually everyone has is a telephone.

Why the telephone? The telephone is a particularly irritating invention. Symptomatic of it's operation is its ability to interrupt some other activity to which you were already engaged. Like that bell in your elementary school, the telephone unapologetically muscles its way into the forefront of your attention, quickly and effortlessly breaking your concentration.

The mission of telecommunications is to enhance communication between individuals. Modern phone technology can do this, and can help to alleviate some of the problems and pitfalls of basic telephone service.

I know what you're thinking. *A book about telephones? How can you possibly spend a whole book talking about telephones?* Well, a prolific writer can spend a whole book discussing any topic. Whether it be bears, beers, or bras (now *that* would be interesting), it's amazing how much can be written about almost anything.

And so, there you have it. This book *is* about telephone technologies, and their application to your not-for-profit group. But, it is also designed to educate the uninitiated, harried individuals who usually make up the "decision makers" in not-for-profit organizations. This book is centered around individuals who are considering installing a small phone system, above and beyond the "ring, ring" you will get from Wal-mart.

Like my last book about computer technology, this is a book for a lot of people: It's for secretaries, "jacks-of-all-trades", volunteers, plant managers and priests. If you answer the phone for a living, manage people who do, or manage the system which is answered, you will find benefit from this book. If you work in the industry, you probably will be familiar with most, or all of the concepts in this book. It is a book for technology committees and home-school associations who have decided to open their doors to new avenues and opportunities in technology application.

Foremost, this book is a first step in a long series between the vision of increased efficacy, and the signing of a contract with someone who will actually sell to you and install for you a phone system.

There are many people who have far more experience than myself, and could write a much better technical book. There are also many people who could write a much better book about not-for-profit organizational structure. This book is neither a technical tome, nor a process manual for not-for-profits. It is merely a book that can help to solidify and clarify concepts about small telephone systems for small organizations.

This book is about telecommunications. Business telephone systems, PBXs, and other communications technologies, and how they can help your not-for-profit organization do business more effectively. I hope that you enjoy reading it as much as I enjoyed writing it.

Joe Liberto
March, 2004

Section 1:
The Changing Role Of Telecommunications In Not-For-Profit Organizations

This book leads us with a question: "Why do not-for-profit organizations exist"? Those who are knowledgeable in the realm of societal behavior say that such factors as "market failure", or "government failure" lead the necessity of the not-for-profit organization. These scholars pose that there are certain kinds of services which cannot financially be oriented to be handled effectively by the government or the private sector. These activities must be handled by interested volunteer effort, such as the quintessential example of the volunteer fire department. Fire fighting is, by nature, not a good thing for a market economy. The equipment is expensive, and who wants competition in this field. The government would not be able to handle this, on a large scale, effectively. Fire fighting would become slow and bureaucratic. Not good qualities for such an activity.

There is a not-for-profit organization to address most all aspects of modern social life and humanistic need. There are groups which address issues in the areas of religious health, education, physical and mental health, social services, civil rights advocacy, art and culture, recreation, the environment, to mention but a few. Whether you refer to these organizations as being the independent sector, the civic sector, non-profit, not-for-profit, charitable, or philanthropic, they are a force and a presence in American society.

What kind of force? According to a recent study, in 1996 there were approximately 1.5 million registered tax exempt (501 (c) (3), and 510 (c) (4)) organizations in the United States. The Merrick school of Business at the University of Baltimore showed a quickly growing economic force in the Independent sector. They shows that, as a percentage of the local economy, [the equivalent of] 18 percent of the full-time jobs in Baltimore are in the Not-for-profit, or Independent, sector.

But what does this really mean? As I formulate this part of the book, I am watching the funeral of America's 40th president, Ronald Reagan, live on the television. Several things occur to me. First of all, America loves it's leaders, for better or for worse, and it goes without saying that we honor those who have served this country in their time of death as in life.

Reagan was different, though, from many of the leaders whom we have had, at least in the relatively short time in which I have been alive (I'm 26). His optimism, and fervent belief in the abilities and, more importantly, the potential of

3

America and the people of this country was, and still is inspiring to me as a citizen of this country. To a degree shared by very few leaders, his views and beliefs colored the tone of the entire decade of the eighties, in a positive way. It colored the message of our music, our clothing styles, and our ideas about religion and on being American. I truly think that we would be different now, and maybe not as enlightened, had we had a different leadership through this decade. I also think that, to a significant extent, the goals to which we are all trying to achieve as those in the independent sector were embodied by this remarkable leader. I say this not as a republican or a democrat. Partisan politics aside, there are certain themes, I believe, that are always seem to recur when people describe Reagan, whether these people are public officials on television, or ordinary people who remembered the time of his presidency.

Those of us who are partaking in educational initiatives, such as reading a book about bettering our sector, are, probably, are more likely to be idealistic and hopeful about the future. We believe in the potential of our volunteers as people and the potential of our client base. We believe that if we can do better, that our client base will live better, happier lives.

I believe in this, myself. I am working on a steering committee which is comprised of people in the north Baltimore county area. Primarily, we are people who work and live on the York road corridor, north of Towson. Our name is Abraham's Tent Coalition (If you would like to help, you can visit us on the web at *www.abrahamstentcoalition.org*).

Our vision is to bring services to those who are experiencing homelessness along the York road corridor. Our mission is to bring gateway services to those who have fallen through the cracks, primarily due to a lack of infrastructure in the area where they happen to be living.

The coalition's plan was going full force for two years, until we got involved with the county government. A primary requirement, and core obstruction to the realization of our goal was a physical facility, and the appropriate zoning and legal issues surrounding the establishment of this location. We quickly found that we were unable to raise sufficient resources ourselves to begin leasing, or building/ buying space to hours our program. We realized that in order to find success, that we would be advised to partner with a larger organization. Initially, we partnered with St. Vincent de Paul of Baltimore. St. Vincent de Paul helps a great many initiatives flourish. They began collaborating with Baltimore county government, trying to get funding for this project.

The county worked well with us, and with St. Vincent de Paul, and we began to make progress. However, they abruptly, and significantly made a change to our vision which has splintered and changed the direction of our fundamental vision. Namely, they moved our project site to a facility which they already own on Rossville boulevard, in Essex. For anyone familiar with Essex, it is nowhere near York road, and is not located in the North of Baltimore County.

Instead of rolling over, we regrouped, and revised our mission. We have grown, and have moved forward. After a great deal of soul searching, we felt that our goals are still as valiant, whether on York road or on Rossville boulevard, and our volunteers strong. Even though we have lost some support, I think that we are stronger today than we were six months ago, and that we truly are moving towards a greater good.

I think that the real triumph of the independent sector lies in the vision and action of the people of a community. People who care about their communities, and believe that with help, their fellow men (and women) will be able to pull themselves up, and better themselves and their lives, are the basis for our sector. In her book, *"Non-Profit & Government: Collaboration and Conflict")*, Elizabeth T. Boris describes this as "social capital". Social capital links people to their communities, and when this kind of linkage occurs, communities, and society at large is improved.

Non-profit organizations, regardless of purpose, create communications networks amongst people. The independent sector, at least as much as the governmental sector, or the private sector, is about successful relationships. People walk into assistance centers across the country, and are served due to successful communication, planning, and good relationships.

One aspect of good relationships is good relationships with volunteers. Volunteers are usually those who have some skill, or expertise or perspective on life which they can offer to the greater purpose of the organization. This perspective may be a fervor, or a particular skill such as communications ability or the ability to mold technology to improve communications. Technology is a hot-button topic in today's non-profit. Communication has always been a hot-button issue.

While discerning this, I have been going through growth in my personal life. This past year, I bought a home. I had been renting for a long time, and over the summer I finally got bitten by that bug. You know the one I mean, the one that decides that what color the walls are *does indeed matter*, and that it is finally time to mow one's own lawn and tend one's own garden.

So, what was the first thing that I did when I moved in? Well, the first thing that I did was to repaint everything. Then I installed wire for a telephone system. It was a fun project, and really got me in the mood to start working on a lot of other things which needed to be done. It was also an education not only in putting together small telecommunications systems in general, but also in the unique pleasure of residential construction.

Even before I had found a house, I had the phones. I bought a used AT&T Merlin phone system at a swap meet for $75, and a friend and I pulled wire through the walls of my 70's townhouse. A great deal of sweat and cursing later, we were able to piece together a really cool four-extension system. There is nothing like intercom for a home. I think that this was the first inclination that I had to write this book.

My personal interest and expertise has been more with education and the not-for-profit sector, so this book is naturally geared towards the kinds of grassroots organizations who have old, creaky houses as offices, and who need this kind of perspective on applying technology.

So great, now you're looking at this book, thinking that I'm going to spend 100 pages talking to you about how to solder thing or something. But, no. This book isn't a how-to guide, to be sure. This book is really a tool, an instrument to spark awareness amongst a group of people who may need this type of technology more than anyone. Many larger businesses look at telecommunications technology as a "given". It is the cost of doing business, a necessity. It's like computers, but more pervasive in a "general public" kind of way.

It has been my experience that non-profits do not have any "givens". Consider these two facts from Elizabeth T. Boris' recent book on non-profit and government relationships: *Collaboration and Conflict*:

In the last five years (from 1994–1999):

* Total private contributions of money and assets have fallen as a percent of income
* The rate of contribution amongst higher-income individuals has fallen

The future, in the independent sector, is not a given. As with any resource, telecommunications and office equipment comes to organizations in different ways. Especially smaller non-profits, often have to rely on those swap meet specials, or the benevolence of neighbors for their "luxuries", such as computers or photocopiers. At the very least, this book may help to answer some questions about the options and directions available to you.

In an era of uncertainty in government and corporate support, and individual contribution, increasingly efficient communication is important to the success of carrying out an organizations' mission. Teachers who cannot get the technology in their classrooms to function correctly cannot teach their children. Parishes and congregations are key environments where communication is necessary. Needing to reach a priest is a highly personal and urgent need. Not being able to do so can work against good will. For any for-profit company, or independent organization which depends on funding or the benevolence of community, the necessity of maintaining goodwill cannot be overstated.

What is communication? There are several varieties and settings for communications. There are personal communications, which exist between individual people. Personal communication is often spontaneous, and informal. This type of communication can range from purely social, to business-oriented, and thus perhaps more thoughtful and serious.

Personal communication occurs all of the time on the telephone. Most of the time, people are trying to get in touch with each other, and find it more useful to make the call, rather than to drive to the location. Utilizing telephone technology to facilitate personal communication is of great benefit.

There is group communication. Group communication involves three to five people. The purpose of the group may be social, or in terms of a work-group, or other task oriented setting.

Business telecommunications technologies are making the conference, or group communication setting, a more positive and effective experience. Rather than try to schedule a location, and coordinate three to five schedules, a group conference call can quickly resolve misunderstandings, or attend to decision-making without having to find common ground.

Finally, there is public speaking, which is not particularly useful for the purposes of this book, but it is highly organized and formal. Usually, it is scripted, and some pre-determined message or agenda is intended to be trans-communicated to a larger audience. Occasionally, if your system is in a school, you may find yourself using the Public-address functionality to make a "speech" to your audience. At our K-8, our principal says a prayer, and reads announcements every morning, using Public-Address technology. As a leadership function, it schedules the organized and pervasive presence of the leader (the principal) into the daily lives and minds of the children.

Just as there are different settings for communication, there are also different functions within these settings. There is the social function. Interaction without an obviously stated agenda. The social function builds goodwill, when used at an organizational level, and works to build other "social capital" by encouraging networking of like-minded individuals.

There is also the decision-making function of communication. This is where there is an obvious purpose and agenda, which is the predicate for the conversation. Unlike purely social conversation, there is an understood goal behind the conversation, to come to an agreement, or to make a decision of some sort.

How can the power of technology be leveraged to help communication in your organization? There are countless ways in which communication technologies can be applied to (1) increasing efficiency and ability to communicate with the outside world, and (2) smooth out small issues before they become real problems due to lags in internal communication. The following case study about a k-8 school should help to illustrate this further. Schools are a great example, because they show the extreme of two elements. The first element is of a staff whom cannot leave their posts. School faculty are assigned to be in certain locations, based on scheduling prepared at an administrative level. Faculty cannot "walk down the hall" to find out information. Taking the time to make the phone call may be a hindrance to their ability to perform their duties.

The second, intertwined issue, as you will no doubt see in this example, is that the ability to quickly resolve internal problems is key to the efficacy of the educational mission, and the usability of the time. I worked for a university a few years ago, and I remember having a discussion with the head of the technology support department. The key priority was the classroom, because the classroom was the "production environment". If there was a technology issue in a classroom, people were not getting what they paid for. The primary mission of the university was to educate, and as such, the primary priority, above administration or faculty office computers, was the equipment in the classroom. Whether your organization educates as a primary mission, or helps in some other way, the area of convergence between your technology application and your primary mission should be your priority.

St. Timothy's School

The day at St. Timothy's begins at 8 A.M., to be exact. Sister Agatha has a clock on the wall to ensure that this is the case.

By 8, she has already checked her voicemail. Three old messages, which she saved from yesterday, and two new messages. She writes the numbers down on a pad, and deletes all of the messages.

Sister Agatha clears her throat, and presses the "all call" button on her telephone, a sleek digital telephone. The LCD screen says "ALL-CALL". Her phone then beeps, as do all of the telephones at St. Timothy's, and she begins the day in the same way that she has every day for the last 22 years, with a prayer.

The technology has changed, but it has not changed the message. She used to walk out to the hall, and flip a series of switches on a large wooden console. As the lights would light up, beeps and buzzes would be heard throughout the building. The old system was loud, but not really reliable. More often then not, Ron, the facilities man, would have to repair some loose wire, or a speaker gone bad. Announcements, especially in the latter days of the PA system, were a daily adventure at St. Timothy's.

She gives a warm welcome to the students, and mentions the weather, and the clothing drive on Wednesday. She reminds the students that fall will soon become winter, and to remember to bring in warm clothing.

After the announcement, she receives a call from a parent, asking about the interim grade reports. During the 5 minute call, two other calls come into the system. Both are fielded by the secretary in the office. One is a parent, whose call is sent to the principal's voicemail box directly, the other is a cold-call from a textbook salesman, whose call is sent to a general mailbox for "pest" callers.

The general mailbox is checked by the secretary, who sends the follow-up messages directly to those who may be pertinent. She can do this easily, because all of the mailboxes in the system are housed on the same computer, and are all interconnected.

Meanwhile, a light blinks on Sr. Agatha's telephone set, indicating that she has a new message.

It's 9:10 A.M. on a Monday at Room D of St. Timothy's K through 8, and that means Spanish class. Miss Smoot is trying to load a videocassette tape about the one man's travels and experiences with the culture of Peru. Miss Smoot knows Spanish, but she doesn't know a whole lot about videotapes. Somehow, the tape has jammed in the VCR. She calls the office, but the secretary and the principal have no answers.

Where is Ron? They try in vein to call his office in the basement, but to no avail. He is nowhere to be found. They try the offices, other classrooms. Room D is turning into pandemonium, fifteen minutes have passed.

In a school setting, the primary goal and mission is the education of the children. Classrooms are the factories, the production environment. If the equipment has failed, it is necessary to locate the people who can fix the problem, and quickly.

Miss Smoot remembers…Intercom/All-Call! She picks up the receiver, and presses the "All-Call" button. All of the extensions in each classroom and office emit a long BEEP, she apologizes for the interruption, and calls Ron, the technology guru to room D.

Ron arrives within a few minutes, and with a wave of his hand (and an assortment of screwdrivers) fixes the errant videocassette tape.

The children cheer, Miss Smoot swoons, and Ron "the hero" bows out to a round of applause. (Well, maybe this is a bit of an exaggeration, but things do return to normal) and thanks to the quick communications ability of an integrated telephone system, a valuable class is not wasted.

Ron will receive six calls that day. He will be in his office a total of 15 minutes. In the days before the telephone system, people at St. Timothy's used to leave him an assortment of notes written on scraps of paper, and on the backs of napkins. "Fix the light in my room", or "The sink is dripping in the second floor boys' bathroom". He would pick these messages up, lose half of them, and wonder which of the fifteen rooms the light was not working in.

Of the six calls, three of the calls were from internal extensions, and three are from outside of the phone system. Ron spent an hour recording the outside message, which introduces him as the "director of facilities at St. Timothy's", and invites them to leave him a message with their name and phone number after the tone. For the internal callers (those whom he works with), the voicemail system will read them a "special" internal message, which tells them that Ron is going on vacation next week. The voicemail system will also play for Ron the name of the caller (for the internal callers, only), and the time that the message was left (for all internal and public callers).

Ron can quickly press a button to reply to their message, telling them when he will be around to check out their problem.

Since he has been given voicemail, Ron needs not get any more written messages. People can call his extension, and if he is not there, they can leave him a message. If Ron doesn't pick up the message within a short period of time, the voicemail system will automatically pick up a regular phone line, and will call Ron's Cellular telephone. When he picks up the phone call, the automatic voice will tell him that he has a message. Although this feature can be scheduled with St. Timothy's system to work only during work hours, Ron lives close to the school, and thus leaves it on all the time.

Last weekend, Ron got a call on his voicemail that the glass had broken in a window. Instead of walking into a mess on Monday, he was able to get to the school and clean up the mess, and patch the hole. Because he doesn't like surprises, Ron is happy with this situation.

That very afternoon at 3:15, school announcements will be given from the convenience of the principal's office using that very same system. First, a group page to all of the phones will be given, with the day's birthdays and a short prayer or a statement from the principal.

To help facilitate the orderly flow of people to dismissal and pick-up, following this general announcement will be announcements to pre-programmed "paging zones", which tell certain blocks or groups of classrooms that it is OK to dismiss.

After the announcements, a voice message will go out from the school nurse, using a group messaging function, to announce that all of the signed "lice information forms" need to be returned to the nurse's office by the end of the week.

Fr. Joe receives a call in his residence from an old friend in Toledo. Toledo is a long way from here, and Fr. Joe likes to call his friends and family. Even though St. Timothy's has five telephone lines which are interconnected to the telephone system, Fr. Joe never uses them. He has a button on his telephone which pulls up his private phone line every time, and no other. In this way, he knows that he will not be making personal long-distance calls on the regular lines, and can keep the information separate.

Now, it's Monday night, around 11, and the offices of St. Timothy's Parish have been closed since 5. Nobody is around, save Fr. Joe, who is getting ready to go to sleep.

Normally, during the day, when someone calls the main number for the parish or the school, they receive the day message, which welcomes them to the parish, and gives them the school (press 1), the religious education department (press 2), and the parish office (press 3).

In the afternoon, when the last secretary leaves her shift, she presses a button on her phone to switch the phone system to from day mode to night mode. Night mode has a different auto-attendant setup, with different welcome messages. The night message tells the caller that the office is closed, but that if it is an emergency, they can reach a priest by pressing 1. They can also get basic information, like office hours, directions, and a schedule of ordinary mass times from the pre-recorded messages on the menu.

If they press one, the call is automatically routed to Fr. Joe's telephone in the residence. It also rings the phone in his office (in another building).

This ability to take a call from a night mode gives the impression that the caller can receive service 24 hours a day without having a whole host of evening staff on call to manage phone calls.

Now, this book really isn't about schools. Having this in mind, though, a school is a great example of an organization which can really leverage this type of technology to release the burden off of overworked support staff. I have seen too many private schools whose administrative staff are mired in phone calls, and are unable to complete other work.

So you see, in the daily activity of a very ordinary and conceivable—and very small-scale—institution, the voicemail and automatic features of electronic telephony are used heavily and are relied upon. There are very conceivable and very appropriate uses of this technology. These are but a few ways to significantly improve the quality of not only the jobs, but of the interpersonal working relationships of people in these types of organizations.

Even in today's vast electronic jungle gym of e-mail, networks, et cetera, telephones are inherently a comfortable and natural medium for interpersonal communication. Telephones are "instant-on" devices, which when well-planned, are very easy to use. This may be the essence of the difference between electronic mail and telephony. Telephony requires little thought-overhead be expended, and is quickly understood and recognized. The interface (the beeps and buzzes) are very straightforward and familiar to most people. The human voice, too, is much less formal than the written word. Because of this, voice transmittal continues to be accepted as a welcomed and appreciated medium.

What does your organization do? Do you help people who need food or shelter or housing? Does your organization teach, or help children understand what it means to vote?

All over this vast country, benevolent organizations are fulfilling a need. The specific nature of the need varies from organization to organization, but the basic idea is the same: The improvement of society at large. Some organizations help to improve the life of lots of people, some organizations are focused on helping one individual at a time.

Some organizations are, like the YMCA or United Way, thousands of people, both paid and volunteer. Other small organizations, like the United Churches Assistance Network (UCAN) in Cockeysville, Maryland, are ten or fifteen volunteers using an extra couple of rooms on the premise of a generous Catholic church.

What do these organizations have to do with you and your organization? Well, their missions and visions may vary significantly from ours, but we all share certain basic similarities. Beneath our missions, we all require a certain amount of space. We all require electricity, and comfort heating and cooling. We all talk to each other, and need to be able to talk to the outside world. Most all of us require communications technology to some degree in fulfilling our missions.

Communications Technology. What do those words mean to you? Do you immediately think about Facsimile machines, modems, or cellular phones? Does it mean calling plans and on-peak/off-peak calling times. Well, Yes, but there is a great deal more to telecommunications than the services offered by the telephone company.

Only the smallest organization may find that a single phone line and a fax machine are the extent of their needs. Other organizations are struggling with the communications maze. Some find that this ordinary arrangement is lacking in some way that somehow inhibits the organization's staff, but they do not feel that they have enough funding or justification to improve their telecommunications position. These organizations look jealously at other larger organizations who have seemingly "high-end" features, like multi-line phones, main-menus, and voicemail, while they can't get straightforward information from the phone company about their monthly bill.

There are a myriad of myths about telephony. Some of these myths stem from a lack of understanding of the state of the technology today. Some come from the historical shroud of obfuscation and uncertainty propagated by the Bell System.

Unlike the business telephone situation under the monopoly of the [Phone] Company in the 1980s and earlier, in today's telephony market there is a basic truth about communications technology. Increasingly, today and as we progress into the future, there is less which is technologically out-of-reach for a small firm.

Just as computer technology blossomed under increased popularity and competition, the global market has opened up to telephony. There are many companies offering a wide range of technologies, and systems for different types of markets. There are a few specialized features which may be somewhat expensive or unnecessary for your firm, but most features have been integrated into very convenient and inexpensive packages which can be operated and administered, if not installed, at very low cost and inconvenience.

While it would not do a service to you to feature or recommend branded merchandise, the likely size and scope of your organization has been taken into account with regard to this book. The author would presume that if your particular organization has a technology position, that you would not likely be reading a book such as this. With this premise in mind, it is this author's desire to present solutions which presume that your firm not have a full-time "technology guru" on staff, or a large technology budget.

Unfortunately, there is also another great truth about small business, both for-profit and non-profit alike. Most organizations are operating commercial-level ventures with residential-oriented technologies. This is largely due to cost and availability. "Big box" stores are selling to a certain kind of client. They are selling reasonably easy-to-use and install, and reasonably inexpensive equipment. Equipment which was designed to operate in an environment where a significant amount of "down time" may be acceptable. As such, too many small organizations are not leveraging the real potential of technology.

Whether we are dealing with phone systems of printers, things are changing in the electronics industry daily. Currently, however, most residential telephone systems are not any more advanced than they were fifty years ago, save the innovation of pushbutton dialing. This is to be expected, because the residential telephone network in the United States is remarkably simple, and hasn't fundamentally changed since electronic switching in the 60's. Most of the changes which have occurred have been centered around equipment complexity and cost, and increased call volume management. Anyone who has a cellular telephone has more technology at their disposal, in a more convenient package, than they typically have at home.

You may find that your clients call and get busy-signals, or long wait times. People call, and are put on the "black hole" of "hold", while an overworked receptionist juggles other callers, live clients and other responsibilities. People recount information to those "on the line", acting as an unnecessary link between what is going on in the office and the person on the other end of the phone link. Call volume and communications deficiencies are exceedingly difficult to detect and diagnose. Most of the time, the only effective way to detect a problem is reactively, by hearing about some call-problem from the client. Many prospective clients or prospective contacts may be lost in the interim between the advent of a problem and the epiphany of that problem, either by proactive detection or reactively by the report of a caller.

How can telecommunications technologies benefit your organization? Effective implementation of multi-line, multi-extension phone systems can have a marked impact on the efficacy and efficiency of support staff. When implemented correctly, auto-attendant systems, which employ menus and call-routing can improve communications, and reduce the overall workload of a "receptionist" position. Some of these systems can eliminate the need altogether for dedicated telephone answering personnel.

Voice-mail, in it's most basic form (the answering machine) has been around for decades. Many people view answering systems with scorn, because they are fundamentally opposed to talking to a machine. I can understand this feeling.

There is a local furniture company in Cockeysville, Maryland, whose stock and trade is commercial furniture and office furnishings. The first thing which strikes a caller to this company is that there is no auto-attendant. A real secretary greets you every time. There are four secretaries, who double as greeters for walk-ins. These ladies (they're always ladies), greet you with a consistently competent and pleasant demeanor. There is also no voicemail. Should you wish to leave a message for a salesman, or for the president of the company, a message is taken by the pleasant greeter. Whereas there are some companies that one would feel uncomfortable, this particular firm pulls it off with panache.

Panache? A word you don't see in the world of the not-for-profit. However, having some style is important, and being effective and seeming competent and positive is important, too. The nature of the telephone strips away so much in the way of both verbal nuance and non-verbal communication, that the veneer becomes very important. First impressions do, in fact, count for a great deal.

Apple computer is a great example of a company who understands panache. In the 1980's, Apple had live switchboard operators. They provided the operators with business cards, and generally treated them with a level of respect commensurate with any professional position. In fact, they referred to this position as being a "Customer Relations professional".

Apple understood that for callers, the customer relations person *was* Apple. For customers with questions or problems with an Apple product, Apple Computer is not Steve Jobs, CEO, it's John Smith, customer-relations professional. Repeat business and satisfaction hinges on these kinds of details. How do your customers see your organization?

This is not intended to steer you away from auto-attendant or voicemail systems. Should you decide to use it, voicemail can have a great many benefits. These powerful systems offer increased privacy over centralized "answering machine" style messaging systems. Voicemail can give an appearance of consistency and professionalism to an organization's public face, as well as a much-improved management and control strategy to incoming messages.

Speakerphone, one of the most desirable and useful features of a commercial-grade telecommunications system, can allow an organization to have conferencing ability with clients and staff, and thus eliminate that "link" from the office staff and the third-party.

EDUCATION

What does your organization look like? Are you in an old house, or in the church offices? Does your organization have office or utility space?

What does your organization look like to those who have never been there at all? The face of your organization is through your phones. Nothing is more counter-productive than calling an organization and facing a labyrinth of menus, or the voice of a hostile bureaucracy: The voice of a tired, overworked attendant.

Whenever someone calls your organization, they will be greeted by certain non-verbal cues. How many times does the phone ring when someone calls. Is the phone ringing twice one time, ten times the next. Consistency is important with your outside callers.

When the phone is answered, is it answered by a live person, or a main menu. Main menu technology can be of great help, but your organization may be providing the kinds of services where this kind of technology application can be a harbinger to your success.

If your organization answers the phone with a live individual, what is the tone of his or her voice?

Yesterday, I went to a different supermarket than I usually do—I needed to buy some milk, and I didn't feel like going to the regular store, so I went to the one on the way home.

Customer relationships, no matter what your line of business, should be the first and foremost priority. I was struck by one thing in the store. I felt like a distraction to the check-out clerk. Here I was, milk in hand, and the clerk is talking to her friend, who is standing on the other side of the checkout aisle. The clerk barely makes any contact with me, other than to ask if the card I'm using is a credit card or a debit card.

There was a time when one would be shocked and amazed by this rude treatment. I was neither shocked, amazed, not even surprised by it. I do, however, identify this kind of treatment now with Food Lion grocery stores. It may be an unfair association, but it's there in my mind, and it will most certainly affect whether I return to that store again in the future.

Your organization, too, has a personality and a face. On the phone, intonation is the subtle clue as to the meaning behind the words. In many instances, voice tone is much more important than what is actually said. In the first spoken word, the attendant has conveyed to the caller what level of importance the caller is to the organization, and the feeling that the attendant has about the organization. Are your attendants screaming their dissatisfactions to everyone who calls your organization? Are they crying for help, because they're overworked, an the phone is a distraction.

Pretty heavy stuff, eh? Is the caller of primary importance, or are they a distraction from some other task. The setting of the phone system should be in congruence to the attendants. Loud ringers, and a distracting "grand central station-esque" environment should be avoided. Above all, the caller should feel that they have called a calm, organized environment where things are running smoothly. Of course, this sometimes cannot occur, because your organization may well be anything but organized and calm. Many are not.

Getting beyond "hello", is the next step in the process. Are the attendants and staff listening to the callers. If you have called a support line recently, you may have found that you have to enter many key pieces of information, such as your phone number, and the model or serial number of the item with which you need assistance. You spend the time to do this, only to find that none of this information is available to the attendant, so you have to repeat it again.

Recently, I called the support line for Hewlett-Packard about a problem that I was having with a client's computer. I called the line, entered a lot of information about the computer, using automated menus, only to have to repeat this information again to the first-level attendant. I spoke with two other people, and had to repeat at least some of this same information to both of them. It was clear that if there was some kind of automated call-management system, that it was not being utilized to it's greatest advantage.

Your callers, whether you have 1 employee or 300, should be treated as customers. Try to inconvenience them as little as possible. You are familiar with how your organization works. They, likely, are not familiar with the structure and process of things.

Hearing their requests and needs, and sometimes their complaints, and follow-up, like keeping notes on such things, is the mark of a good listener, and a good attendant.

FUNCTIONAL TRAINING

There are two types of training for phone system users. There is the functional training which accompanies any complex office equipment. "Press this key to make this happen, don't unplug that because something might go awry".

That kind of training should include several levels of individuals:

1. System managers
2. Attendants
3. Ordinary users

System managers are the people with the keys, so to speak. These individuals should be trained to a level which may exceed the scope of this book. These individuals should know how to add people to the voicemail system, modify the auto-attendant menus, make moves, adds, and changes to the telephone switch. If possible, one person who is technically handy should invest in at least a few of the tools specified in the "basic tool kit". Probably the most important tools are the 4-pair tester, and the Toner-wand.

The system manager is going to be the tier-one support for your system. This person needs to know where the equipment is, must have a key to the system closet. If there is a service agreement in place, this person should be the contact to make the judgment-call about calling in the service. The system manager is just that, the manager of the system. Give this one person (or small group of individuals) the responsibility over the system, and don't allow everyone to make changes on their own.

Conversely, the attendant should be taught all of the user features of the system. He should be taught how to transfer and put calls on hold, how to log into the voicemail system. This person is the "guru" of the system. This individual has a great stake in the success of the phone system, as it will link to their ability to be successful at their own job.

The users should be given an appropriate amount of training to utilize the features which you have given them. Even in a small scale organization, getting users together to train is a difficult feat. Most users need be taught only basic features, such as:

* How to transfer a call
* How to conference others into the call
* How to put a call on hold
* How to adjust the volumes of ring, handset
* How to check voicemail
* How to delete voicemail
* How to leave messages for others in the system
* How to use other key features which you intend for them to use

Users can be given a "cheat sheet" which covers all basic features of your system. This can be beneficial, and will encourage them to use the phones to their full extent.

Make sure that people are familiar with the correct use of these features, as there is nothing more irritating or confusing to a caller than to be shunted around the phone system, or worse, hung-up on by someone who cannot successfully transfer a phone call.

SOFT SKILLS

We spoke of the importance of being able to converse effectively with the caller. Soft skills training is primarily for the benefit of the attendant(s) of your system, although effective telephone communications skills are important for all of your staff. The goal is to improve communications within your organization, and with your outside callers.

Focus on soft-skills, like:

* Call screening: Which calls are important, which are not. Many attendants simply cannot screen calls. They simply receive calls, and become a transit. If the caller is knowledgeable about the organization, the attendant is merely another step in the process of getting in touch with the real destination. If the caller is not familiar with the staff of the organization, the attendant can be a real monkey-wrench in the lines of communication.

Train your attendants in the way the organization is set up. These are really your knowledge workers. Invite them to be a part of your organizational debriefing, no matter how small-scale it may be. Your attendants should be treated as part of your staff, not lowly functionaries.

Keeping attendants in the loop on who does what is a great way to get your clients to the right people, on the first try.

If your organization is small, or if you are blessed to have a private secretary, develop a series of call lists to keep callers organized. You can develop a black-list, for those callers with whom you never wish to communicate. There should also be a white-list which includes select people who call irregularly, but should always be allowed to get through.

Think that this is a draconian way to handle people who call? Well, the phone is primarily a tool for your organization's benefits. If you look at the phone as a tool for your productivity and as a time-saving device, you will find that your view of caller—rights versus staff—rights may be clarified.

* The endless transferring of callers, and staff interruption can be minimized by helping your attendants to get basic information to the caller in a succinct and accurate manner. Instill the idea that the attendant is responsible for the caller. The attendant should be made aware of news items and informational issues which may cause someone to call the organization. This can be aided by keeping information sheets in close proximity to the attendant operator. How can the

operator successfully *help the caller*? This should always be in the forefront of the operator's mind.

More importantly, there should be lists of information which should not be given to outside callers. Cellular phone numbers, home numbers, or private information may be undesirable to give out.

Especially as of late, there may be reason why the media would contact your organization. Your attendant is rarely the individual whom should be speaking for the organization. There should be a clear understanding of this fact on the part of the attendant.

* In the event that the problem cannot be solved immediately, gathering basic facts from the caller, such as:

- *a name, a phonetic spelling if necessary:* If the person is calling on behalf of another individual, or of a company, this should be recorded as well.
- *a call-back number, and a good time to return the call*
- *When, and on what date, did the caller call*
- *Concerning what issue, or what information or service that they require*

There are a few others, which are equally as important, yet very often not used:

- *Promised call-backs:* Did the attendant, or your secretary, promise that you would call them back. There is no worse evil than promising a call-back within a certain period of time, then defaulting on that promise. It is also good to note some scale of how urgent or routine the call may be.
- *Point of Last Contact:* If this is a repeat call from someone who is trying to accomplish a task, or resolve a problem, who did the caller last speak with in the organization. Tying this knot between your staff people can greatly aid in avoiding repeated effort, and miscommunication.
- *Impressions of the caller, including tone and emotion:* Is the caller calm or irate. Is the caller dishonest. Listening for cues about the tone of voice in the caller, or the kinds of things which they are asking about, or for, is important to giving a clear impression of the situation to the next-in-line.
* Rather than putting the call on hold for an extended period of time, When can the caller be called back, do so. If possible, schedule the call to be returned at a specific time or day and time.

* Avoid phone tag: This is a tip for all of your users. As the attendant, don't accept, "I'll call back later", or some such nebulous statement. Get a time, and day when your staff can return the call to the individual. This goes into greater organizational topics, such as staff-secretarial relationships, but the secretary should always be kept abreast as to basic work schedules of the staff. This can greatly smooth operations, and gives callers a feeling of synchronicity and efficiency within your organization.

As the caller, don't tell someone that you will call them back "later". Try to pin down a time when the individual will be able to accept your call. If the call will be extended, or of a lengthy nature, try to schedule the call with the staff secretary, so that the individual will be aware of the impending business, and will be able to be prepared for your call.

* Be truthful and forthright. Don't keep too many secrets from the caller. If John is on vacation for a week, tell the caller this in a straightforward manner. Don't let the caller call three times, only to be put into John's voicemail. It makes John look uncaring, and ultimately makes the organization look inept when John does return, and tells the caller that he has been in Bermuda for the last week.

* Setting priorities: The base of success in one's job is centered on effective time management and prioritization of tasks. We are all multi-taskers, especially in the small not-for-profits, where everyone wears many hats. Just as there are jobs which we would rather not do, there are calls that we would rather make. It is human nature to substitute the personal preference for the real priority, which can have a serious negative effect on job success. If someone calls with a request, how urgent is the request. How much priority should it be given over what task was occurring prior to the phone call.

* Try to be efficient. Keeping an eye on the time, keeping conversations brief and businesslike can help to speed along your customer requests. Ask your staff to analyze how long has the average call lasted. Also, how long are people generally put on hold. How many, out of ten callers, are put on hold, not including transfers to other extensions. A phone call is a dance, the goal is not to leave your partner hanging!

Many of these things seem obvious, but they really aren't. A lot of people do not have phone training, formal or otherwise. There is a learned skill to being able to deal with callers. Soft factors like the tone of voice, male voice or female voice (callers tend to respond better, in general to female voices. This is why virtually all voicemail systems use a female voice). A friendly and upbeat attendant can't hurt your organization's public image, either.

Like a lot of things, telephone systems can be of great use, or can be greatly useless. Like any office equipment, spending the money and buying the stuff is 10% of the journey.

The good news is that there is great reward in successful training, and those rewards can be on both sides of the fence. Both your staff, and for your client base, will find that better telephone communication works to their advantage, and makes operation more efficient.

DETERMINING YOUR NEEDS: A WHOLISTIC APPROACH

Any good system begins with planning. One of the worst things about modern society is probably the lack of planning in urban and rural areas. Single-family homes, big-box stores, and shopping centers litter the landscape, because civic planners cannot, or will not, make the decisions necessary to keep things organized and thought out.

Massively expensive homes are built, and are not planned effectively. This results in massive disasters, both for builders and a line of hapless owners for years to come. Lack of planning and forethought leads to ugly structures, ugly systems, and a general feeling of apathy and low satisfaction.

Defining your needs is the first step. Talk with your attendants. What is good about your system, what is bad about it? How many users will you have? Who will be the first-line point of contact?

The physical design of the workstation is important, too. Are the phones to be placed on the left of the desk, the right of the desk. Place your phone jacks accordingly with your furniture plan. The furniture plan should drive the location of data and phone jacks, and power outlets. NOT THE OTHER WAY AROUND!

The physical placement of the phone is also of importance. Is the phone hiding under a stack of papers or books? Is there a place for the phone, or does it perch where there is room?

If your users are right handed, or left handed, makes a difference also. Being left handed, and in the minority, has made me appreciate this dilemma at a personal level. I prefer the phone to the left, so that I can pick it up with my left hand. Most "righties" are opposite to this. The phone (and by association, the desk) should be corded in such a way to allow placement of the computer, mouse, and telephone in such a way as to be natural for the user.

If you have a credenza, or a cabinet of some sort, this can be an ideal location for your technology equipment. Computers, and phones can be placed out of the way of your primary desk space, keeping everything organized.

Near to the telephone should be several other convenience items, such as a notepad and pen. Also, there should be three other things: The internal phone directory, the cheat-sheet, and the actual telephone manual (given to you with the phone).

Installing any telecommunications (or data communications) system should start with a comprehensive site survey. Typically, you will have already selected your physical location. Even if you are planning to install your system into a location where you already are housed, the steps are the same.

What is a site survey? A site survey is an investigation of the building. You should note several important factors. Most of these factors have to do with the installation of cabling. Other factors have to do with finding a suitable location for a central wiring point in your office, such as proximity to the desired locations for phone stations and computer equipment, and the presence of or potential for reliable and adequate electrical service. There are many questions which should be addressed in a site survey. Here are a few key points. These, by all means, may not be the only specific issues which you encounter. All buildings differ, both in nature and in structure, and the issues of your particular building may not be addressed here.

The first rule of a site survey is to pay close attention to the physical surroundings. Attention to physical details and construction nuances is key to a successful site survey.

It has been my experience that taking a pad of paper and a camera to a site survey can be very helpful. Take photographs of things like the ceilings or walls within the building. Draw a picture of the layout of your offices, and the locations of your phones. The phones must be located near access points which avail themselves to having communications interfaces (wall jacks) installed. If there are key areas which have no walls, under-floor raceways or wireless solutions may be necessary.

Go home, and figure out where things like desks and other furniture will be placed, and what the lay of the land is, so to speak. You will likely be doing this kind of activity anyway to determine your space and furnishing requirements, so this aspect of the site survey should not be too painful or out of the ordinary.

Who Owns the Space? Most businesses, and non-profits alike, do not own the space they inhabit. This is OK, as real ownership may not be the issue. The first issue in a site survey is to determine how invasive you may be in terms of wiring a telecommunications system. Is the "landlord" going to be responsible for this service, or are you going to have to furnish the wiring. Perhaps the landlord has furnished the cable, but it is inappropriate or inadequate for your purposes or your system. Is there an issue with the relocation of wire, or the addition or removal of an older system? This issue, specifically, should be discussed and agreed upon, preferably in writing, prior to the decision to move into a location. This is especially important if you have already determined that you need to install a communications system, because any changes you make may have to be in accordance with the wishes and under the supervision of the landlord or ownership of the property.

Office Layout: Where are the offices located inside of the building? Is the office space on the ground floor, basement, or an upper level?

Do you have access to the wire closets, or the demarcation point for the phone company (Demarcation points and other jargon will be discussed later). Basically, can you get access to the point where your phones connect to the outside world?

Somewhat attached to this idea is the ability to control access to the equipment. You will likely be installing central hardware for your phone system. Phone system hardware is both delicate and expensive. If you are sharing the demarcation point with other tenants in a multi-tenant building, it is important that you can re-locate the main phone line to an adequately secure area within your office space.

Ceiling Type: Note the construction of the ceilings in your location. Is the ceiling made of drywall or a similar gypsum product like the walls? Is it modular (often referred to as "puzzle") tile, or acoustic ("drop") tile. Most modern commercial buildings and remodeled spaces will have drywall wallboard and acoustic tiling for the ceiling. From a wiring perspective, the most difficult environments are drywall or plaster walls and ceilings.

Acoustic, or Puzzle tile, called so because it fits together like one of those sliding number puzzles with the numbers 1 to 9, was very popular before modern acoustic drop tile technologies. It can be a friend or foe to an installer of wire. There is usually a key piece of this puzzle (ha, ha) which can be freed to allow access to the plenum (the air-space above the tile). These key pieces will have a metal tab on the side. Be careful with puzzle-tile, because many of the popular patterns are no longer available, and breaking this kind of tile is exceedingly easy to do.

Walls and Wallboard: Similarly, check your walls. Plaster, Concrete and block walls are difficult, and usually require that surface-mounted applications, like Panduit or Wiremold be used.

Flooring: Just as there are different considerations for walls and ceilings, floors can be of great help, or a nuisance when trying to install wire. Wire application exists for under-carped applications, but it is not recommended for use where there is heavy foot traffic, or otherwise hectic locations. Bear this in mind when determining desk layout.

Wire Closet Location: Wire closets and wiring will be discussed in greater depth later in this book, but you should be cognizant of where you will be locating your main wire closet. Is there ventilation or air-conditioning in this space? Is the space secure, so as to prevent theft or vandalism, or to the many would-be technicians on your staff who will want to "fix" your phone system should it fail at some point.

Is there sufficient electrical service to the space? Wire closets need electricity *and proper grounding.* The equipment is very sensitive to electrical surge, and lightning.

STANDARDS, SAFETY, AND QUALITY

Entertain an analogy, if you please. You and I are taking this pleasant walk together down a path to somewhere. Think of me as a park ranger of sorts who has been down these roads before. As we walk, you notice the flora and fauna. Some of these things I point out to you, some I do not. We have a map at our disposal, and periodically I point down one path or another.

I didn't invent the forest, or grow the trees. I also didn't write the guide or know about every plant which grows here. I am not a cartographer, and as such did not draft the map.

Just as the park ranger is something of an authority within her realm, I am an authority within a certain realm of these systems. There are many books which describe in detail the points which are touched on in this book. Books which talk about proper types of grounding, and fire-stopping, and many other topics.

This book is a collection of pointers, so that you become aware that there are indeed concerns out there. Hopefully, it won't be the only, or the last one you read about phone systems for your organization.

Codes and Standards

Codes and Standards are implemented, and are important for two reasons: Life Safety, and Quality of Installation. This is the primary purpose, to protect life, health, and property, and to ensure quality in construction.

Codes are not, of course, limited to telecommunications systems. Codes having to do with wiring began with the advent of electricity, over a century ago. In the US, we have the National Electrical Code, which is managed by the NFPA, the National Fire Protection Association. There is also the Underwriters' Laboratory (UL) which certifies safety features built into electrical and electronic equipment having to do with fire prevention.

All around the world, there are similar electrical standards institutes and regulatory bodies, for instance the Canadian Electrical Code (CEC), in Canada.

According to BICSI, "standards are established as a basis to compare, measure, or judge capacity, quantity, value, quality, performance, limits, and interoperability". Standards are defined by organizations which are [necessarily] not affiliated with particular vendors.

Telecommunications and electronics, in the United States, Canada, and Europe, are highly orchestrated and standardized. The telecommunications and electrical contractors will be cognizant of the standards, and what they mean for your installation. You need be only aware that these standards exist, and in a rudimentary way, how to locate groups who are familiar with them, should any question or concern arise during the installation of your system.

Standards originate in different bodies and committees. Some standards have to do with compatibility. EIA/TIA defines standards for labeling wire so that future people who do service or change the system can easily deduce the structure of a system. A lot of "amateur" telephone installers are not familiar, or disregard this standard. Labeling is very important, and should not be disregarded.

Other standards have to do with personal safety and the prevention of loss of life, and as an incidental factor, the prevention of damage or destruction to property.

Standards usually guide several things having to do with the installation of a phone system. The first is appropriate grounding of the equipment, for the protection of the equipment itself, as well as the protection of the individuals and the building which uses and houses the telephone system.

The second is the fashion and method of the running of cable within a building. Even buildings with wireless telecommunications and networking have some type of cable system, and this cabling system is subject to local electrical codes,

codes set forth by the National Fire Protection Association, and by association the National Electrical Code.

These codes have to do with distances between data cable and other structural supports and things which run in the ceilings or walls of buildings. There are particular standards which govern the type of wire which may run in the building. Some wire, called "riser cable" is chemically designed not to burn as fast. Other wire, called "plenum cable" is designed not to give off as noxious a fume if it catches on fire. Plenum cable, and riser cable are more expensive than non-plenum cable, and thus it is important to ensure that your telecommunications installer is actually using the correct type of cable for the job.

These rules are designed to protect the other aspects of the infrastructure. For instance, you cannot put a hole in the wall for the purpose of running wire or pipes without using a type of product called "fire-stop" to fill the excess space between the wire sheath and the edge of the hole.

To protect the other wires and pipes, you cannot use electrical wire which runs in your building as a support for your telecommunications wire. This means that using wire ties to tie your wire to the metal-clad electrical cable is out.

I mention here several organizations which are of consequence. The first two are BICSI and the American National Standards Institute. BICSI stands for the Building Industry and Construction Service International. BICSI is critical to the telecommunications industry worldwide, because their field of expertise is telecommunications systems. They certify hundreds of individuals annually to install, and design telecommunications systems, and work with telephone companies and independent communications professionals worldwide to ensure that an acceptable level of quality, and more importantly, safety, is present in structured cabling installations.

ANSI, and BICSI do not define standards in telecommunications systems, but they do, however, act as an administrative management center and clearinghouse for information about such standards as proper grounding, proper wire handling and termination (phone jacks), and proper fire-stopping. Fire stopping is crucial, because life-safety is involved. There is a section in the "Infrastructure" chapter regarding fire-stopping practices.

Standards are defined by groups like EIA (Electronics Industry Association)/ TIA (Telecommunications Industry Association), who certify and test electronic equipment, and define standards for communications hardware and accessories.

This brings us to another organization, called the IEEE. The IEEE defines standards having to do with quality of installation and standardization. Phone systems are like those "Lego" toys you played with as a child. Think about how irritating it would be if there weren't standards in the production of those toys. Some of the holes would be far apart, some closer together, and all would range in size. It wouldn't be a whole lot of fun.

Finally, we have the NFPA. The NFPA is the National Fire Prevention Association. They define standards having to do with preventing fires from starting. Because electrical and voice/data communications systems can conceivably cause fires, the NFPA has some things to say about how it is installed in a commercial location.

The Key ideas:

* NEC codes have to do with safety in materials usage and installation technique
* IEEE, TIA/EIA define standards which have to do with interoperability, compatibility, and quality
* BICSI does not define standards
* ANSI does not define standards, they administrate US standards from other sources.

If you are interested in the codes which relate to communications wiring installation, there are several which are important.

NATIONAL ELECTRICAL CODE: *(available through the NFPA, and electrical supply houses)*
* **Section 90.2** : Provides information about what is covered in the NEC
* **Article 100, part I** : Defines terms such as bonding, ground, premises wiring, and signaling circuit
* **Section 110.26** : Defines space requirements for working clearance around electrical equipment. This is very useful when planning space requirements for telecommunications equipment closets
* **Article 250** : Grounding information
* **Chapter 3** : Discusses wiring methods, including raceways
* **Article 645** : Discusses information on equipment, power supply wiring, interconnection wiring, and grounding of IT equipment.
* **Articles 770 and 800**: IMPORTANT chapters about optical fiber cables, and communications circuit codes.

<u>ANSI/EIA/TIA Standards of Interest :</u> *(Available through BICSI)*

* **568-B.2** : Defines standards for balanced Twisted-Pair Cable, 4-pair and multi-pair cable (specifies amount of twist required for telephone cable grade certification)

* **568-B.1** : Cable requirements such as distance, outlets and wall-connectors, physical topologies, field testers, and minimum transmission requirements for different types of communications wire.

* **570-A** : Discusses requirements for residential building cabling system installations.

* **606-A** : Discusses requirements for labels, color-codes, record-keeping, and grounding and bonding.

DEFINING YOUR "ON-PAPER" TELECOMMUNICATIONS NEEDS:

How many staff do you have?
What is the Layout of your office space?
Is there a Wiring Structure in place?
Do you need weather-resistant or specialized equipment?

Have you ever surveyed your internal staff, or volunteers? One of the key steps to determining needs is to ask questions. Your telephone attendants will be only too happy to share with you the stories of trauma and frustration caused by your phone system.

Look at it from a different perspective, and write down your notes. What if you were interviewing your secretary about her desk, what kinds of questions would you ask him or her?

* Is there enough work space, or do you feel cramped?
* Is there a place for everything to be stored so that it is readily available?
* Is the desk comfortable in terms of ergonomics?
* Does the computer fit well on the desk?
* Do the drawer slide mechanisms work consistently?
* Do the drawers slide in and out easily, even when full?
* Do the locks work consistently well?
* Is the appearance and functionality holding up well?
* Would you buy this desk if it were your decision?
* If you could change any one thing about the desk, what would it be?

These kinds of questions, while structured, get people thinking about the nature of the object, and its functionality. While the questions should be asked, invite other comments and tangents as well. The purpose of the discussion is to listen to your staff, and to gauge their reactions, and to take into account their concerns. While there are other factors to consider, the primary purpose for a communication system is to facilitate effective communication with regard to the staff of your organization. This factor really should have the greatest amount of value placed on it.

Observe people's body language, and other nuances about how they answer the questions that you pose to them. Sometimes, these factors are more crucial than what is actually being said to you.

Invite your staff, also, to a group discussion to discuss the above talking points. Explain your desire to find out the shortcomings of the existing system, but focus more on the dreams and "wish-lists" for the future. Write everything down, or tape the group dynamic. It is out of group listening sessions that good ideas arise, sometimes good ideas which have nothing to do with the task at hand.

These are some good starting points to develop a needs document for your organization. A good brainstorming session with all of your staff is a beneficial strategy. Get people together, and talk about the types of work you do.

Some organizations will find need in certain types of features. I worked for a small organization who saw clients who were experiencing trouble with meeting expenses, and needed sheltering. Sometimes, clients came in because they had trouble maintaining transitional (temporary) and permanent housing situations. I watched a few interviews, and realized an immediate symptom of a problem.

Every client, new or repeat, would get a chance to have an interview with a volunteer. Here's how the interview worked: The organization had desks set up in a main central office which were manned by volunteers. The purpose of this organization, in part, was conflict resolution between clients and third-parties, such as creditors or the gas and electric company. This organization also acted as a gateway to certain other benevolent services, such as clothing and food assistance programs.

At the onset of the interview, the volunteer would take the client's information, and have a conversation with the client about their current situation. Based on the information given, the volunteer would explain to the client what assistance services they may have need for, and may be available. They would then converse with a variety of different parties on the phone, depending on the nature of the need.

Two things struck me as being counter-intuitive about the situation. The first was that the volunteer had to recount most of what was happening to the client, and then to the person on the other end of the line.

The second was that there was a list (handwritten) of about thirty-five telephone numbers to call for various people in different government and benevolent organizations. The list was broken down by topic and region. The whole place seemed to run on sticky-notes.

It seemed to me that not only could time have been saved, but additionally, a higher degree of service afforded to the client had the organization used a telecommunications system with (1) speakerphone, and (2) Computer Telephony Integration, in the form of an automated auto-dialer. As we will see, speakerphone and autodial are not distant or unattainable features.

The real issue is that the planners for this organization never thought about the limitations of the telephones, and their impact on the speed and accuracy of the volunteers. Reasonably, this group was all too happy to take what came to them with the space which has been given for their use. Instead of looking at ways to improve the system, they accepted what was available.

But wait! You may be thinking that your entire budget for technology, as is the case with theirs, may be a three- or four-digit affair. You may have blown your budget for technology on this book. However, advance planning can make your recruitment of donations immensely more effective.

For instance, in Baltimore we have a benevolent organization called the Knott Foundation. Knott, amongst many other things, funds technology advancement in benevolent organizations. In a recent seminar on grants and funding which I attended, a trustee of the Knott foundation was speaking about applying for money for technology projects. The first priority of a grant for technology is to explain, succinctly, the needs of the organization, the size, and the scope of the project to be funded. Because the window of opportunity for funding (and other donations) are relatively small as compared to the window of opportunity for a traditional purchase and installation, it is much more effective to have written down the scope and needs in some form prior to the application process.

Your telecommunications plan, or maybe your wish list, should start with a plan for the layout of the offices and/or workspaces, and a list of the staff. Telecommunications systems tend to work better when there is a one-person to one-telephone relationship, even if the phones are not always used by the same individuals.

Especially when considering more advanced features such as personalized caller-id (for internal extensions), dial-by-name, phone call auditing and voice-mail, the 1 to 1 relationship is critical.

While considering the number of staff and offices, consider how the phones will connect to the system. Desks should be placed somewhere around walls which can be accessed to place wire and communications jacks. If you find that your organization has a "Main desk" which is not wired, or similar arrangements, a wide variety of other options are available, such as hardware installed within the floor, or under the carpeting. There are even unsightly poles manufactured that can extend from the desk to the ceiling, which can be configured to carry electrical and data cabling.

The next consideration should be the existing system. By the existing system, I may be referring to the existing wire. Wire is very important to a telephone system. It is the most expensive, and hardest to replace, factor in any installation. In practical terms, you may be working a phone system around the existing cabling. Hopefully, you can adapt the existing cabling to the new system, at least for long distances, and/or hard-to-pull-to locations.

Dealing with legacy technology, and processes, is one of the greatest challenges which face network operations people everywhere.

There is an overwhelming urge, which is too often reinforced by many, to treat legacy equipment in a cavalier, "slash-and-burn" fashion.

In my opinion, dealing with technology in this way can be wasteful of resources which have already been put forward, and disrespectful to the time and talent of the previous system manager/technology committee. The key to success is to try to disconnect your own feeling to want/need ownership over a project versus the greater good of the organization.

If possible when dealing with a legacy, that being a pre-existing telecommunication system, the first thing to do is to have a long and honest conversation with the previous administrator of the system.

Telecommunications equipment, especially digital PBX equipment, has distinct personality. This personality pervades everything from the automated "voice" which outsiders hear on the phone, to the way buttons are placed and labeled on the telephone sets.

Because of the continuous necessity of telecommunications, it is more important to provide your users stability. Make certain that the system is working, and in place, prior to a move. Make certain also that if you are moving into a new space, that adequate time has been given to teach the key people how to do certain things, like

* Transfer calls
* Dial internal extensions
* Put calls on hold
* Make outgoing calls
* Access voicemail (If you have this feature)

More tips on this will be described in the section on "telephone directories, and user training".

Legacy equipment can be more than the phones on the desk, however. One of the first telecommunications experiences that I had at St. Joseph Parish was the installation of a new Public Address system. Yes, PA systems are very similar to phone systems.

We investigated the existing wiring in our school (which was 2 wires, and a metallic shield, to each room). Specified in the plan was the idea of creating a bi-directional system, whereby the teacher could speak to the console by pushing a switch next to the speaker-box (located in the office). Alternatively, the office could speak to one, a group of, or all of the areas where a combination speaker-box/microphone was located. The office console could initiate an "intrusion", whereby they could initiate a listening session into the area where the speaker-box was located.

Upon installation (which started on August the first), we found that with the old system (made by AIPhone) required one pair (2 wires) to each speaker. The new system, however, which was manufactured by Bogen, required four wires to the speaker to operate the switch, three of which were required to operate the speaker and microphone.

We had only two wires to each of the 28 locations in our three separate buildings.

So, what to do? We could not afford to pull new wire, and the old system had gone to the trash, so that was out. Well, here we all are, sitting in my office, trying to pull together some kind of solution, and it's the sales rep and I and the installer. I'm playing with a piece of the speaker wire, wrapping it around my finger, and pulling apart the insulator, and you can see the light bulb flash on over the installer's head. We'll use the shield as the third conductor.

It truly isn't pretty, but it works, and they have 2/3 of the solution. Is this a success? Well, no, not really. You can have the bi-directional ability, but it can't be initiated by the remote, only by the console.

At the risk of alienating the people at our sound system contractor, the problem was, and the real lesson is that you shouldn't rely on the judgment of the contractor alone. In our particular case, we simply assumed that the new system would work with the same wiring pattern as the one being replaced. It occurred to no one of us that there would be a difference until we began doing the installation. Because they specifically stated that the old wire would be used for the installation, when drafting the contract, we were faced with a significant increase in the cost of the system should we pull new wire.

Like our public address system, unless the building is truly "new", there is likely to be some form of legacy telecommunications technology in place. I have found that this is the key difference (today) between telecommunications and data communications systems. There is usually the luxury of being the "first in line" with a data communications system, because usually there is none.

There is most likely one of the following wiring situations in the space you do, or will inhabit. This includes ordinary 2-pair or 3-pair (4-wire, 6 wire respectively) residential-style communications wire, 25-pair station cable from an old business-style system, or 4-pair category 3,4, or 5 grade data/voice communications cable.

What is the most useful?

Aside from a usual telephone arrangement, 2-pair basic phone wire is useless. The last time this wire was useable for a network installation was in the 1980s, and the ability to run PBX-type phone systems on it is eliminated not so much by the wire itself, but by the way it is almost always installed. In most installations, you do not have the "home-run" wiring to get 4 wires exclusively to a location for an extension, because jacks are daisy-chained (hooked together, one to the next) throughout the premise.

This is not to say that this arrangement is completely useless. In many small non-profits, modems and facsimile machines are used for other connectivity. These devices work very well with this arrangement, and can co-exist with your phone system. In this situation, plan on the communications system requiring the installation of new wire.

25-pair station cable, which is historically either tan, gray, or "Bell System" green, is a thick (like heavy electrical wire) bundle of wire which snakes around offices. Many times it was stapled to the baseboard, and has been painted over at least a few times. It ends up somewhere in the room, and has either a plastic wall-mounted box, or an open connector attached to the end. Do you have boxes? The connector is inside the box! The connector looks like an aluminum harmonica, and is typically referred to as a "Centronics 25-pair" connector. Some people call them "Amphenol" connectors, because of the brand-name of a manufacturer who used to make them.

These cabling systems can be remarkably useable. As most of the original systems required a "home-run" wiring scheme, there is usually a pile of wire running through the building back to a central location. Both the older PBX and key system arrangements which used this type of wire required electricity at the phone closet, so there is likely an outlet handy.

How can you utilize this technology? Electronics stores used to sell a device which was relatively inexpensive (under $30) which could mate with the 50-pin connector on most phone systems installed by Bell in the 60's and 70's, and allow you to adapt the 50-pin connector to a set of 6 separate RJ-45 (4-pair Ethernet-style) jacks. The 25-pair wire (50 wires) can be broken into six groups of eight (6 x 8 = 48, plus two extra).

Well, perhaps you have wire hanging out, but it has no connector of any kind. Someone lopped off the aluminum harmonica when they made off with the old phones! People will do strange things, like take those "cool looking" phone-system telephones home with them, only to find that they won't work without the rest of the system. (I actually removed a system from a doctor's office once, and the doctor had taken the proprietary telephone home, because he wanted to be able to have speakerphone at home! But I digress).

In some cases, and depending on your particular phone system, there is a possibility that you may still be able to use the wire. Especially if cost is of concern, this may be preferable to having to pull new wire. Bear in mind that because there is no significant amount of pair-twist in 25-pair station cable of this type cannot be used for computer networking purposes, and that even certain newer (digital) PBX systems will not work with this type of wire. For more information about this, see the section in the book about "Attenuation, cross-talk, and loss". Basically, when dealing with multi-pair wire, *Twist matters*.

This brings us to the third kind of wire that you may find. If you have been blessed to move into a newer building, or a remodel, you may find that you have the right infrastructure already.

If you have faceplates with those "extra large phone connector" (RJ-45) jacks, you likely have what you need for many of the phones out there. While you may still have some level of adaptation at the head-end of your phone system, you are likely ahead of the game.

As you may have gathered, the goal is to adapt whatever you have to the RJ-11 (4-wire or 6-wire) or the RJ-45 (8-wire) jack. These are the backbone of the telecommunications arsenal. Virtually all modern phone systems use one or the other to connect desk telephones to the backbone wiring.

Where do we go from here?

To review, Once you have determined a need for a telecommunications system, there are three steps to begin your journey:
• An initial site survey, including the basic layout and nature of your office space, your electrical service, location of wiring closet.
• A needs analysis, of some sort, to determine the level of system that you need to install
• A Follow-up survey (wiring paths, existing wiring, and station locations)

What will follow?

• Meeting with system integrators or those who may provide you with equipment or services
• Talking with your local phone company's business office to discuss features and pricing
• Installation and training of your staff

PRIVATE BRANCH EXCHANGE, KEY SYSTEM AND MULTI-LINE TELEPHONES

Types of Telephone Systems vary significantly. There is a range of systems out there. Some may come to your organization through donations, some purchased new. There are mechanical telephone systems which you may have, or run across which were commonly installed by the Bell System in the 1970's and 1980's. These systems, called Key Systems, are still very viable for small firms, and equally as reliable as today's electronic systems. Because of the longevity and sturdiness of these systems, many have been allowed to remain installed in older buildings and continue to service today's clients as they did when they were brand new.

There are several common classifications of telephone Systems. The most basic are single-line and multi-line telephone systems. These are basically ordinary telephones, some with built-in answering machines.

SINGLE AND MULTI-LINE PHONES:

Here it is, folks. The majority of existing systems in the United States are what is referred to as "POTS", which is the first acronym. POTS, which stands for Plain Old Telephone Service, is a tongue-in-cheek acronym which has become normalized over time into regular use in the communications industry.

POTS telephone service is very simple and rudimentary. This simplicity begins at the level of the infrastructure. In most small buildings, especially those built in the computer-networking dark ages, telephone wire is strung from jack to jack, in a daisy-chained fashion. For instance, lets say you have the first floor of a townhouse. You have a front sitting room, and two smaller rooms in the back which have been converted to offices.

Take a look at the phone pole. Where is it in proximity to the house? For our example, let's say the pole is in the front of the house. There is a real probability that the telephone wire enters the house at the front, and was fished through the wall to the closest phone jack to the front of the house.

OK. So what about the other jacks? Well, in most converted residential homes, a second wire would be connected to the first phone jack (in the front room), and would fish through the wall to the next closest wall jack, in the first office, for instance. From the phone jack in the second office, yet a third piece of wire would connect the third phone jack in the third office to the second phone jack in the second office. This would continue throughout the series of rooms in the building where it was presumed that only one tenant would occupy the phone system.

There is a fundamental problem with this paradigm. In a purely residential environment, it would be unnecessary to isolate individual telephones from each other. In fact, other than in the bedroom of the phone-bound teenager, it would likely be undesirable to make only one phone ring, because you would have to run down the stairs or into the other room to pick up the phone. Because homes don't have receptionists, there is no call routing, and thus no need for wire isolation. Daisy-chaining works just fine.

In a small commercial environment, this is a real limitation for the exact same reason. Meetings may occur in an office, or people may be conducting business on the phone which is private or important. Because even most modern telephones do not have line-in-use indicators, there would likely be a continuous problem of picking up in-use lines, which is just what happens in a home where there are more than a few residents.

Multi-line phones are not interconnected or "aware", as such, of the network other than through the normal means of the telephone system. This is the normal way that phones occur in the home/residential environment.

Feature Attributes:

* Unless defeated, all phones ring when a call is received
* One cannot call internally from one extension to another
* (Usually) there is no built-in indicator that a line is in use
* (Usually) there is no hold or music-on-hold feature
* Stations are usually daisy-chained one to another, with no individual "home-run" wiring or a central wiring location necessary.
* No control or auditing functionality is built-in
* No auto-attendant functionality is built-in
* This is the default type of system which will be installed by the telephone company if no other is specified by the owner or tenant of a building.

KEY SYSTEMS

What does it mean to be a Key System? Key telephone systems mean simply that the phone sets (the part that holds the handset) has other buttons ("keys") which are separate in function from the dialing buttons. Possibly, functions like "Hold" or "Flash" would be exceptions. Usually, the key functions are to either select other lines, or to select other internal extensions within your system.

In this book, a key system means that the phone has buttons which, to one degree or another, are programmable in function. Usually, these buttons will select other outbound telephone lines, or dial other extensions within your local phone system.

The key system is the next step beyond a single- or multi-line phone system. Key systems really became popular in the mid 1970's with the introduction of the Com-Key 416 telephone system. For smaller businesses who needed more features than were available on the truly featureless residential phone systems of the era. To say that these were "full featured" systems is a misnomer, as the feature set is rudimentary.

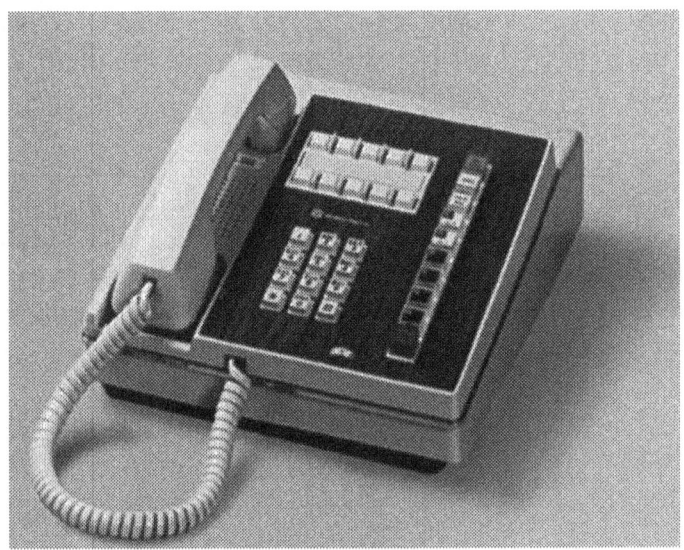

Figure 1-1: Example of a Key-type desk telephone

Early Bell System key systems, which are still very common in older buildings, are electromechanical phone systems. They were devised and built by the Western Electric company for the Bell system, before divestiture in the early 1980's. Older buildings may still have these systems, which may have very iconic telephone stations, with buttons employing real lights and buttons. Some, as the one featured here, utilize solid-state memory, and more advanced features.

Why are these systems important to us? As will become apparent throughout this book, there is a fundamental truth about non-profit organizations: This is that many things are donated. This being said, there is very little likelihood that your organization will receive a donated key system (a true key system, that is). These systems are very bulky, and while not terribly difficult to install, they tend to be hard-wired into the buildings to which they belong, and are very proprietary to the number of lines and extensions to which they were configured.

This is the reason for their inclusion here. While it is unlikely, though not impossible, that your organization would buy and install a system of this type, it is very likely that your organization will find one if you find yourself located in an older building.

What kind of features can you expect to find with a key system? Key systems are, by nature, very simple telecommunications devices. They do not have a great deal of features, but this does not mean that they are not useful to your organization. Most key systems have all of the following features:

* The ability to place someone on hold
* Multi-Line Access
* Extension Signaling (The ability to "buzz" another extension)
* Selective ringing (not all phones ring when an incoming call is received)
* The ability to co-exist very successfully with facsimile machines, modems, and answering machines. These systems work in parallel, and will not integrate with voicemail or other attendant features.

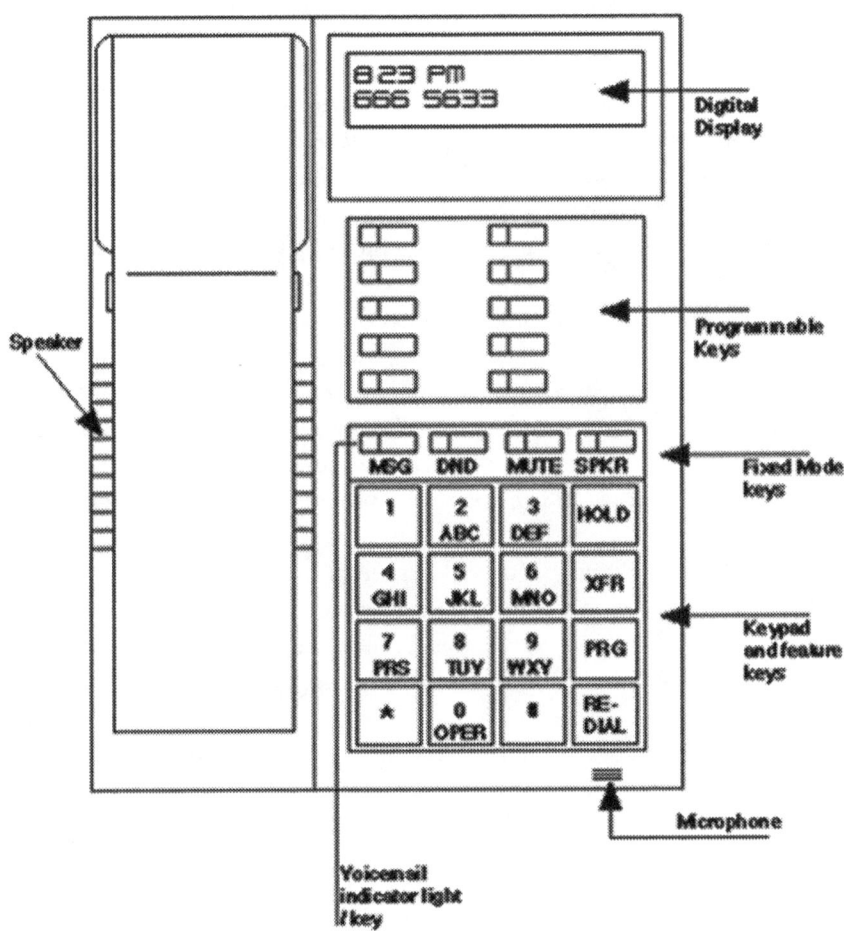

Figure 1-2: Anatomy of a typical PBX Telephone

PRIVATE BRANCH EXCHANGE (PBX)

What is a PBX? PBX stands for Private Branch exchange. The PBX has one requirement, which is that *each extension within the system has a wire which goes directly from that particular extension to the central "hub" of the system.*

In the most modern systems, there are a few proprietary systems which violate this rule, using digital technology, but the 1-wire to 1-station rule is a good rule of thumb and is standard practice within the industry.

PBX systems have a cabinet located somewhere, usually around the area where the phone line comes into your office area, where all of the wires meet. The purpose of these cabinets is to link all of your offices extensions to each other. Extensions will be required to have internal numbers, like numbers 10 to 20. In the most basic system, these numbers are affixed to the phones, and are not changeable. In more advanced systems, additional extensions may be added to the basic system, and all extension numbers are programmable by the person who sets up the system.

Extension numbers allow you to make internal calls, or to intercom between phones without having to have multiple outbound telephone lines to make the connection. Just as you would expect, the "Private" in PBX indicates that this system can operate independently of the telephone company, as a small, internal system, if this is desired.

The cabinet is like a small, automated, operator. Each extension has a wire which connects to a plug on the switch. Using the number keypad on the phone, one can dial other extensions within the system, just as one would dial other phones outside of the premise. Usually, there is an alternate button to dial extensions, versus regular phone numbers. Alternately, it may be required to dial a special number, like "9", to access an outside line prior to placing a call.

These next generation of systems to be introduced to the market were basic multi-extension office systems. These systems are still widely used in small businesses, and are marketed to offices and light industrial applications.

Possibly two of the most the most popular of these systems were the "Merlin" and "Partner" series manufactured by AT&T.

Figure 1-3: Example of a PBX desk telephone

PBX systems of this type usually are set up in families of "series", depending on (1) the number of extensions which can be attached to the system, and (2) the number of outbound lines which can be connected. At least one popular system uses a numbered system for their nomenclature. They may have a model called the "2/08" (pronounced "two-oh-eight"), which means that the system can accommodate 2 phone lines, and 8 stations. This same family of products may include an upgraded model called the 8/10 (the "eight-ten"), which means that eight phone lines can be connected simultaneously, and that ten extensions can be handled. Typically, small systems max out at around 15 or 20 phones, some handle far fewer than this.

Critical is understanding the limitations of the number of phone line- and station capacity. Purchasing a phone switch that accommodates too few stations or lines can be a challenge, because typically these systems cannot be cheaply or easily upgraded to accept more stations or lines.

Understanding PBX systems of this type is important, as this is the first generation of phone systems with useful features for small businesses. These full-featured systems are readily available, not especially expensive, and are easy enough to install. Another factor is this: In this country, and in our area especially, there is a great deal of enthusiasm surrounding the donation of equipment. At St. Joseph's, nary a month goes by that someone isn't trying to donate washing machines, refrigerators, computer equipment, and other types of appliance hard-ware to the parish, either for use by the parish, or for subsequent donation to other charitable organizations.

It is my experience that the most likely candidate of all of these for donation is the small, office-type PBX system. Small businesses tend to install new systems when they move. If a member of your board, or volunteer staff, or a benefactor has a small business, you may be able to get an entire system free of charge.

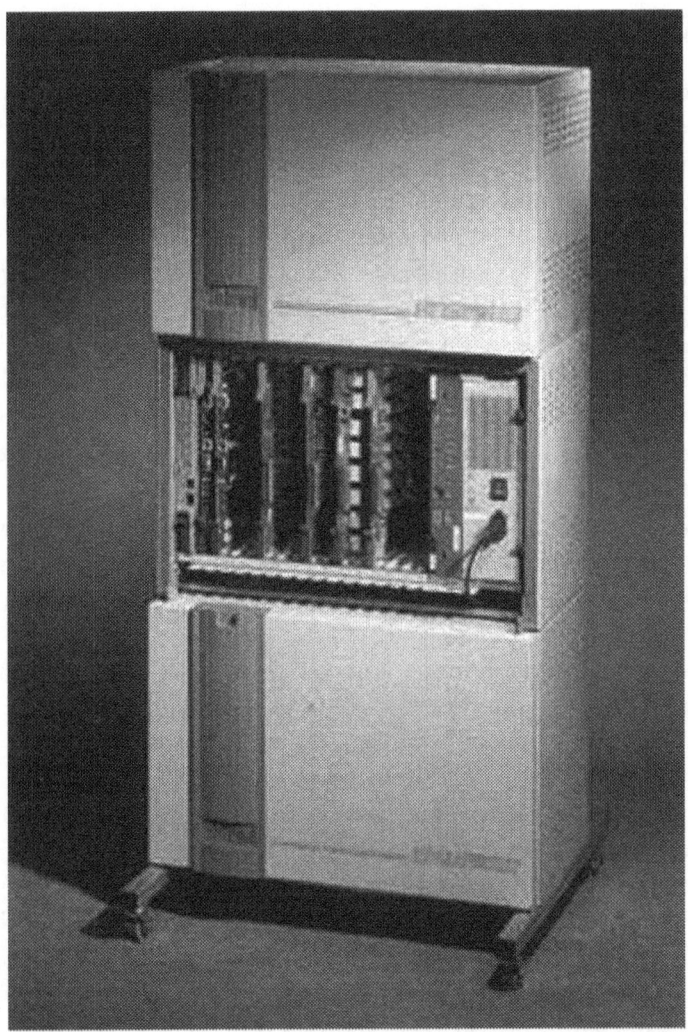

Figure 1-4: Example of a medium-sized phone switch. Note the 3-level stackable cabinets for expansion

ADVANCED PBX SYSTEMS

As time progressed, and more advances were made in the field of electronics, telephone system technology moved forward. Even as far back as the late 1970s, telephone systems no longer employed such antiquated electronics as electromagnetic relays and hard-wired buttons with light bulbs. Light Emitting Diodes, solid-state memories, and other integrated circuit technology were massive advances for phone systems, because they drastically reduced the amount of wiring and the amount of electricity required to run a telephone system.

Consider this: An older system, like the Com-Key desk telephone, in the earlier picture, would require a station cable (the cable running in the building from the cabinet to the telephone desk set) of fifty wires (see the section on infrastructure wiring). The connector looks like a small harmonica. To compare, an ordinary phone line requires 2 wires.

The reason for all of this wire was that each light, and function which was attached to a key required a single set of wires dedicated to that purpose. This limits, inherently, the number of functions which can be attached to a particular telephone set.

Newer systems, however, run on 4, 6 or 8 wires, with small connectors which look like ordinary telephone-type connectors. The wire itself is much less invasive to a building construction, and is a great deal less expensive, and easier to install.

Basically, there are three key differences between the basic and more advanced PBX systems. The first is the number of features built into the system. The second is the expandability of the system.

As the market has progressed, these systems have become increasingly modular. Newer phone system cabinets are basically empty metal boxes with a supply of power of some sort, and a back plane. A back-plane is a large circuit-board with a series of slots, like those found inside of a computer. Whereas a computer would have a modem, video adapter card, or a scanner adapter card plugged into it's expansion slots, telephone systems have cards for music-on-hold, outbound phone lines, and phone stations. Usually, you buy a "card", a ridged plastic module which fits into the metal cabinet, and plugs into the back plane, which accommodates a certain number of extensions. Cards may be available in 8 or 16-station versions. One manufacturer refers to their station cards as being "full cards" (16 stations) and "half cards", which can handle 8 stations. The half-cards are not half the price, but they are markedly cheaper than the full-card price.

Simpler PBX systems, as well as the older systems from AT&T, will not have back planes. They integrate a certain level of functionality into the model, and other than basic modular add-ons for specific kinds of equipment, this is the functionality that is available.

There is a third difference between a basic PBX and an advanced system. IVM, or Integrated Voice Mail, is changing the way companies do business. There is an entire chapter on features of Integrated voicemail, so these will not be repeated here.

Basically, though, the *Integrated* in IVM means that the phones are aware of the voicemail to a degree which is unheard of in an ordinary telephone answering machine situation. The voicemail system can pre-empt the calls, and can transfer calls from the outside world to internal extensions. In the most advanced systems, the auto-attendant (a feature of voicemail) picks up the call on the first ring, presents a series of menu choices to the caller, forwards the call to an extension based on the choices made by the caller, and then retrieves the call when the extension does not answer after a few rings.

IVM gives voice prompts and help, and a "friendly", albeit automated, face to your phone system presence, and can be very effective if managed properly. Basic phone systems very rarely have true IVM functionality.

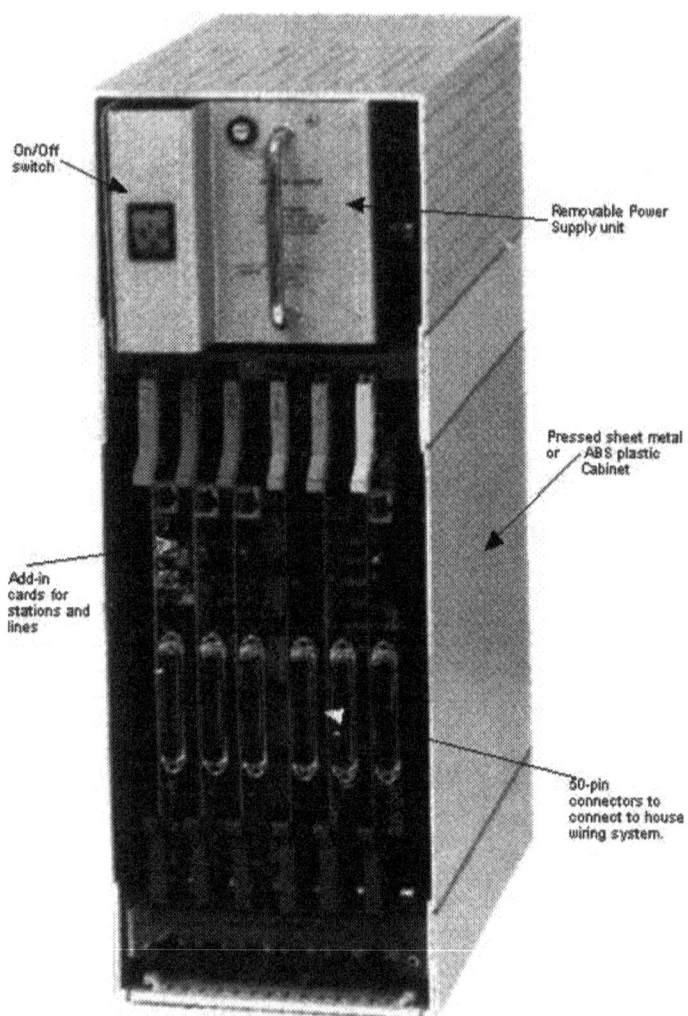

On/Off
switch

Removable Power
Supply unit

Pressed sheet metal
or   ABS plastic
Cabinet

Add-in
cards for
stations and
lines

50-pin
connectors to
connect to house
wiring system.

Figure 1-5: A small-scale phone switch.

DIGITAL AND HYBRID TECHNOLOGIES

Some systems are comprised of a control unit which works using completely digital technology, like a computer. Special integrated electronics utilizing what is referred to as Digital Signal Processor (DSP) technologies allow your voice to be sampled and converted to digital waves. "Sampling" is a technique used in many pedestrian activities, such as the encoding and decoding of audio waves for musical data on audio CD.

Just as the music is encoded onto a compact disc, using a silicon chip to measure the audio waves and record when the music is played, DSP chips in digital telephone systems "listen" to your speech on the phone, they then encode the waves picked up by the microphone into digital streams, and pass these streams through the phone network within your phone system.

DSP's great advantage is it's ability to compress the voice into a smaller package size. Why is this important? Compression frees up the channels within the phone switch for more powerful processing and faster response. It also saves room on the hard disk drives or solid-state memories for message handling on digital voicemail systems.

Hybrid phone systems use a digital system for addressing the basic needs of the phone, such as it's extension, and other management functions. The actual voice, however, is not digitally encoded, but passed through wire as an analog signal, as is the case with the ordinary telephone network. These types of systems do not require as much high-end processing hardware to be included within the telephone switch, and thus can hypothetically be less expensive. To varying degrees, a system can be digital without being completely digital.

As a side note, most all modern voicemail systems are digital, and use compression to save disk or memory space.

From the users' perspective, it will not matter as to the underlying technology of the phone switch. Digital phone systems sound basically the same as hybrid and analog systems. One exception to this is the handling of speakerphone calling. Remarkable advances in speakerphone technology have been progressed further through digital electronics. The newer speakerphone systems are much clearer and pleasant to use than older models.

INFRASTRUCTURE WIRING

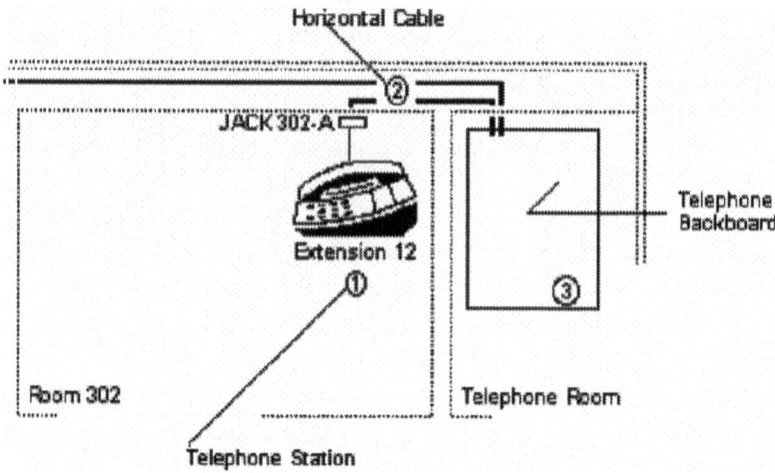

*Figure 1-6: A basic wiring diagram showing Telephone System Backboard
(3), in-wall horizontal cabling (2) and telephone station (1)*

As you have gathered, this is not a technical manual about telecommunications, per se. You will find no diagrams here about how many feet wire can be run and still pass category-3 certification tests. There are volumes written on premise wiring, structure, and certification. Enough technical explanation is being given to allow the reader to understand the basic principles involved in coordinating a phone system installation. Should you need technical information about communications wire, installation hardware and accessories, BICSI, a telecommunications association located in Orlando, Florida, has published a great many compilations of the standards and structure for premise wiring.

That being said, BICSI starts it's courses with a simple statement. The single largest cost factor to new data/telecommunications installations is the wire. It is for this reason that we focus on this important foundational topic in a book such as this. Because many problems can be traced back to bad designs and shoddy installation, it is important that you, the owner of the system, be at least familiar with some of the more critical factors of the infrastructure. The purpose of this section in book such as this one is to kindle an awareness of the wiring schema in your offices. In all likelihood, volunteer or paid contractors will be helping you to run wire in your building.

Telecom contractors are people too, though, and most of us are honest people who want to see the client get the best that they can. There are a few bad apples in any basket, though, and all too often bad apples are not educated in the codes and standards, or are looking to outright cheat the client out of something which should rightfully be there.

So it is with this in mind, that we ask ourselves, What is *Premise, or infrastructure wiring?* Premise wiring is the backbone of your telecommunications system. Earlier in the book we discussed "home-run" wiring systems, versus "daisy-chained" systems. In communications, and to a greater degree data communications networks, there are several ways to connect equipment. These schemas are referred to as "topologies", which means that they are usually defined graphically to show the logical, and sometimes physical attributes of the schema.

Daisy-Chain or Ring Topology:

All devices are connected in series, but this is different in theory from the phones in your home, which technically are connected one to the next. In computer networks, sometimes the series becomes a continuous loop, conceptually, like a circle. If the series, or the loop is broken, some or all of the devices may fail to communicate with each other. Basically, this is an obsolete schema, and is used very rarely in business telecommunications installations.

Star Topology:

Star topologies are the most common type in modern networked systems. Wires run from the end user station, whether that be a printer, computer, or telephone, to a central "Hub". If the link between the end user and the hub is broken, it has no direct effect on other end users' connections to the hub, or to other users within the star.

Think of the hub as the center of the star, and the tangents off of the center hub as the lines of wire running to the to the users. For our examples, the star topology will be implied, unless otherwise specified.

Cross-Connect Fields, and Wire Closets

So, what exists at the outside edge of the star? The telephone desk station, or wall phone, or the door phone. The telephone connects to a connector installed within the wall or the floor, which is in turn connected (on the inside of the wall) to a wire. The wire has been pulled within the wall or the floor through some form of pathway to your central location, or your "closet". Wire closets are very often installed in closets, near to the electrical and phone entry points into a building.

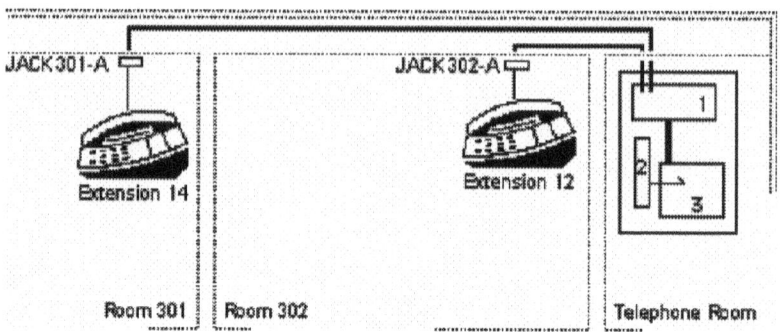

Figure 1-7: A Simple Cross-Connect Block-Diagram showing (1) Patch panel, connecting horizontal in-wall cabling to a connection point, (2) 66-block connecting outside lines to phone system, and (3) telephone switching hardware connected to patch pane (1) and 66-block (2).

In the above diagram, the telephone is connected via a wire which runs through a wall to another room, the wire closet. The wire in the wall connects to the block numbered "1". The "1" block would have wires running to all of the phone jacks in all of the offices. These jacks are numbered with some form of structured system, based on the room number, for instance. Let's say that the room pictured is room 301, so the jack, being the only one in the room, would be jack #301-A. "301-A" would be marked on a label on the block numbered "1" in the picture, so that a technician or installer would know it from the others.

Block number 3 is the electronic phone switch cabinet. It houses all of the electronics which make the phone system work properly. All of the telephone numbers that the system can handle correspond to individual wires which are connected to the switch. These wires connect each extension number to a corresponding place on block 2, which bridges connections from blocks 1 and three.

The purpose of block 2 is to facilitate the management of the telephone system. The second block allows moves, adds, and changes to occur without modifying the existing, organized structure of the extension numbers, which do not change, and the jack numbers, which also do not change.

This arrangement is referred to as a "cross-connect field", and is the model which is used almost exclusively to interconnect and manage telecommunications systems.

TELEPHONE EQUIPMENT CLOSETS

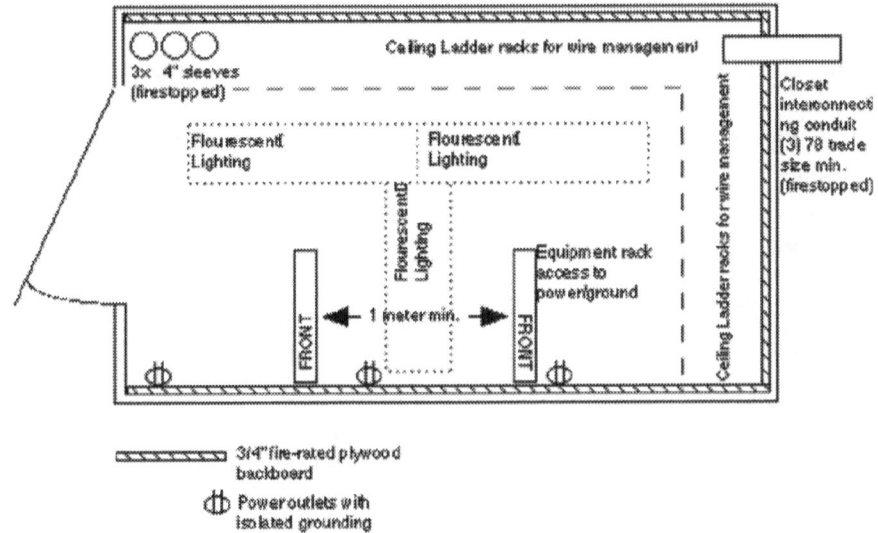

Figure 1-8: EIA/TIA equipment room example

EIA/TIA 569-A defines a "typical equipment room", along with the supplementary wire management, and equipment management hardware. Telephone equipment closets house certain types of phone-system related hardware. They act as the central point in the hub, where all wire must be pulled to from around the building. They act as a location to store and operate the telephone switch, which provides the electronic "central office" function for your phone system.

Additionally the cross-connect field, (the blocks in the previous graphic) will be located in the telephone closet. Phone closets should be set up in such a fashion to allow *easy access* to equipment for maintenance. A phone closet is revisited on a regular basis, to troubleshoot the origins of phone-related problems. You will be visiting this location to execute move, add, and change-orders; that means that this is the place where wires are moved around to change where "extension 10" plugs in. Because Extension 10 is hard wired into port 301-A (Remember?), extension 10 will not function properly unless it is physically plugged into 301-A. Should you wish to move extension 10 to another location, like 308-A (for instance, because Beulah and Meghan can't get along together in the same office), it would be necessary to go to the phone closet, and move the connecting wire which is connected to "extension 10" on block 2 from jack 301-A on block 3 to 308-A on block 3. Now, the circuit is restored, but on a different jack in the wall.

It is good practice to keep a master list of several things having to do with your phone system. Amongst these things would be a list of office numbers, jack numbers, and extension numbers. If your system is small enough, and most are, you should be able to manage the user names along with the extensions. This also makes configuration of their phones easier to manage, and remember for the system manager.

User Name	Office Number	Jack Number	Extension
Chuck Clarke	301	301-A	14
Sandy Green	302	302-A	12
Amy Dumont	303	303-A	15
John Brown	304	304-A	16
Paul Grey	305	305-A	17
Joe Liberto	306	306-A	11
Bob Cook	307	307-A	13
Beulah Armstrong	308	308-A	10

Figure 1-9: A Sample Phone System Management Chart

BACKBOARDS AND DEMARCATION

Looking at the equipment closet in this book is a series of discreet components combined to make a whole. These components range from a simple piece of ¾"-thick, fire-rated plywood, called the Backboard, or Back plane, to the expensive telephone switch unit itself.

The physical back-plane will usually be installed by either the telecommunications contractor or the electrical contractor. The back-plane is important because it provides a workspace for the telecommunications installers, and is the "canvas" for your system. It provides, additionally, the structural support for all, or most, of the following components which comprise your system.

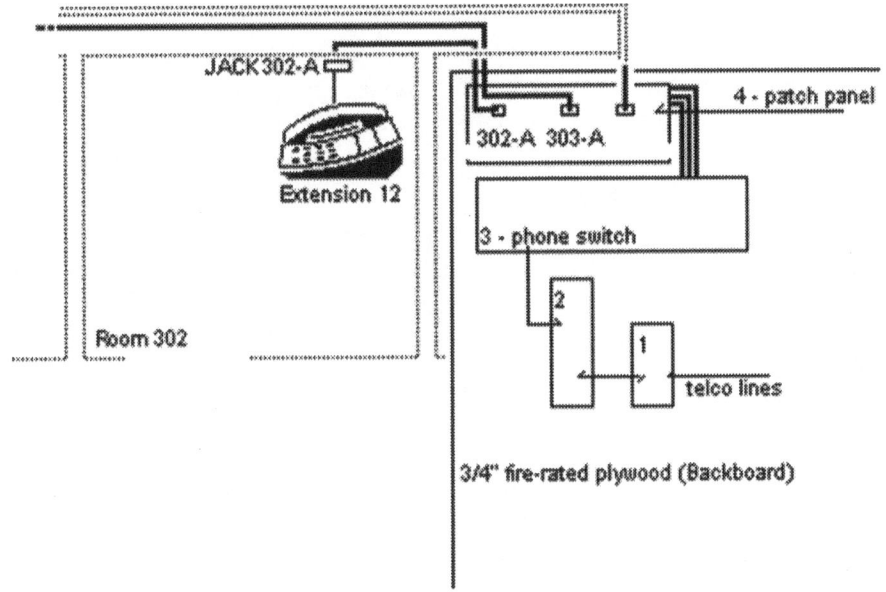

Figure 1-10: Backboard diagram showing:
1. Primary Ground for outside lines
2. Demarcation point for phone lines from telephone company
3. Phone Switch
4. Patch panel connecting phone switch to house wiring

The above picture shows a basic telecommunications backboard with components. Your particular installation may not be in appearance as the one described here, this is OK. However, the basic components (or functional equivalents) should be present in your system.

Let's begin at the beginning. A building plan usually includes a specification for a room or a closet designated as the "Point of Demarcation", or Entry Facility, for the telephone company equipment.

As building progresses, and in due course, the telephone company or the builder's electrical contractor will usually install a backboard. The telephone company will install several components. Amongst them will be a cabinet of either metal or plastic and of varying size. Into this cabinet will be a wire which descends into the floor, or somehow outside of the building. This is, functionally, the last point where the phone company is responsible for the telephone line. This point, called the Demarcation point, or "Demark" in phone parlance, is their junction point to your house-cabling system and your "customer-owned equipment"—your telephone system.

This concept is very important, and confusing for the uninitiated. In the dark ages prior to 1983, telephone equipment was usually installed (and owned) by the local Bell Telephone Company. The phone company installed and maintained not only the connection to the network, but also the physical equipment, all of it. Every desk and wall telephone in every home, office, and otherwise was the property of the Bell System.

After 1983, or thereabouts, and the legal breakup of the Bell Telephone system, there needed to be a conceptual point where the Bell was no longer responsible for the wire, and the physical equipment. This is referred to as the Demarcation point, or entry facility.

Everything beyond that point is yours.

In the 70's and 80's, an industry of "interconnect equipment" service and sales was born out of this legal decision. Most smaller organizations will opt to have an interconnect equipment provider install and maintain the equipment. Your organization may choose to do this as well.

Does this mean that *you* don't need to know about your phone system? *Of course not.* Interconnect companies come and go, sometimes organizations need to move to other vendors. Rule number one in this book is that although you don't need to know every technical concept, you should know the basics of your system.

Beyond the Demarcation point (Labeled 1) is, or should be, a device called the Primary Protector.

Good installation practice says that any wire which enters/exits your building should have grounding and lightning protection devices installed before it physically connects to any of your equipment.

This does not just apply to electrical cable. Your communications cable should be protected, and the phone company's cable should be protected before it connects to your equipment.

NEATNESS COUNTS

There is a universal truth to wiring projects, and activity which endeavors to put wire into a structure. Equipment closets are *not* for extra storage, and neatness always counts. Coming into a room that is full of cartons, extra parts, and clusters of unmarked wire is a daunting and irritating experience, both for the system manager, and the service people.

You will find that neatness has a marked effect on both your ability to manage the system effectively, and your ability to encourage service personnel to come out to help service your equipment. Encourage? Well, service people are like anyone, and there is always something else to do. If you are finding that you have a difficult time getting service on time, take a look at this space. Is it shoehorned into a corner, or behind the boiler in the boiler room?

If rule number one is that you should be cognizant about your phone equipment, rule number two is that everything should be installed and maintained in a "workmanlike" manner.

What does that mean? Being workmanlike is an antiquated term, from the days when Williamsburg wasn't "historic", and goats roamed the streets. It was a term used mainly in contracts involving framers of homes and other buildings and the people paying for the homes to be built.

You have no such agreement with the telephone company. They are obligated, if not by you, by their culture and code to install their equipment to such a level of quality and organization that their service people can repair and maintain it. This is important to Verizon, as they are obligated to service their equipment for the life of the system.

Verizon notwithstanding, You should specify a descriptor such as this in your wiring installation contract with your interconnection system provider, although you may need to word it differently, depending on the company. Wires should be neatly installed, labeled in a logical manner. Additionally, walls, ceilings, and other aspects of the building should be returned to their condition prior to installation.

You should also specify that certain nationally and internationally recognized codes are adhered to. Specifying that EIA/TIA guidelines, and those guidelines which affect your local governance are observed. You should also check all of this yourself.

Wire bundles should be tied together, and should be supported on walls such that they cannot fall, or disconnect from other equipment. There are proper ways to support wire, and a variety of products sold in that industry which are designed for this purpose. Wire should not appear to "hang", or sag from one point to another, this is a clear indicator that there is not adequate support for the weight.

Wire should never be supported by other lengths of wire, and should not be supported by other apparatus, such as the supports for the drop-ceiling. You will need to poke around to see what work has been done. If you see wire that is attached to electrical cables, water pipes, or the supports for drop ceilings, this should be rectified by the interconnect provider.

In any installation, proper supervision is key to success. Watch the people who are installing the network wiring. If you are not knowledgeable, this is OK. Meet frequently with your project's manager from the Interconnect contractor to discuss the work which was done—meeting once a week during the job is not uncommon. Use common sense and ask questions of the project manager, openly and often. Write down the answers to the questions, and then repeat this back to the project manager to make sure that everyone is on the same page. Insist, also, that memorandum confirming the topics discussed in your meetings are faxed to your organization, and kept on file with the contractor.

Take pictures, lots of pictures, of the work being done. Someone in your organization needs to take the initiative, why not you? Draw diagrams of the room, and if things happen where walls are cut open, take photographs or video of the space. Next year, when the contractor is gone, the ceiling will be back in, and nobody will remember where the wire has run. Draw diagrams so that you know where wire runs in the ceiling. If some area needs to be chopped apart to pull wire, try to install an access panel (metal, fire-rated) in that location, rather than simply repairing the ceiling.

Vigilance is important, and keeping the contractor's work supervised is well worth the time, and whatever needs to back-up on the supervisor's plate to do so.

66-BLOCKS, 110-BLOCKS AND PATCH PANELS

Every interconnection system must have building blocks. Traditionally, the basic building blocks of an interconnected telecommunication system are the panels to which wiring is physically connected. These panels, referred to as "blocks", are heavy-plastic modules which screw to the wall. These blocks contain metal contacts which are specifically designed to attach to the small copper wire which comprises most telecommunications cabling systems.

As with any telecommunications product, there are different standards which have been defined, over time, by both Bell System and third-party product developers to handle the orderly interconnection and management of wired connections.

The two most common types of connection hardware are the ubiquitous 66-block, and the 110-type block. There are other standards as well but they are functionally similar enough to one or the other of these as to be disregarded specifically in this discussion.

Why are you concerned with these types of equipment? From a management and users perspective, it is not particular important, as all of these blocks perform the same functions.

There are a few advantages, depending on your particular situation, of the 66-block. There is a corresponding tool, which comprises a blade and a "punch-down" tool, whose purpose is to attach the wire to the block, and to cut the excess wire off, to keep things neat.

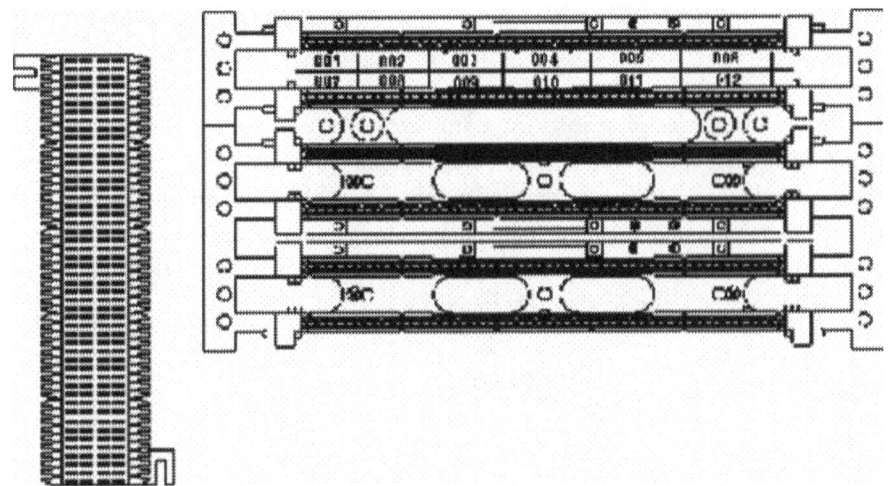

Figure 1-11: Examples of 66- and 110-type punch-down blocks. Left side is 66-block, right is 110-block, showing fronts of both panels. Both types of panels screw to a piece of plywood. Cable from outside of the room comes in through slots on the left and right sides of the mounting hardware, and punches down on the front of the block.

PATCH PANELS

Patch panels, which are functionally different than blocks, perform the same basic purpose but offer increased flexibility and ease-of-change over the block.

There is a great advantage to the patch panel. Patch panels are the modern version of the old "operator desk". The one with the wires and the plugs. Once installed, the panels require no tools to make basic changes, such as re-locating extensions. The wiring system uses normal jacks, like wall jacks, to make connections. Patch panels are called such because they have two sides, a front and back. Once installed by the installer, the back side is more-or-less permanently attached to the wire, going either to the wall jacks or to the extension ports in the phone switch. Usually, the wire will come from all around your offices to this central point, and will all culminate in the back of the patch panel, where it will be installed into the patch panel.

On the front of the panel, there are plugs which are connected to directly to the connectors on the back of the panel. In appearance, they look like the ordinary network/phone jacks on the wall, but they are very close together. Each is labeled with the number of the port in the office.

Basically, the wire runs between the back of this panel and the jack on the wall in the office, and is labeled to identify it on the front of the patch panel.

On the front of the panel, are wires which connect to the telephone switch. Each wire is labeled with a phone extension number (extension 1001, 1002, 1003, etc.). The wires plug into the connectors on the front of the patch panel, which extends them into the jacks on the wall in the office.

Why is this so much better than a punch-down block? It makes moving, adding, and changing extensions much easier. Moving the extensions is a matter of unplugging a wire from one connector, and plugging it into another, then moving the phone physically to the new location.

ELECTRICAL EQUIPMENT

Electrical equipment includes two categories, for the purposes of this book. Equipment directly related to the operation of the telephone system, such as the telephone switch and voicemail component.

The second category is broad, and includes protection equipment, such as surge protection devices, and battery backup devices.

As every phone equipment room will have some manner of electrical equipment, it is advisable to have an electrician install an isolated circuit for your communications equipment.

The isolated circuit would be a single, or group of outlets which are connected to one circuit breaker or fuse in the box, which does not branch off to other devices.

Circuits used for telecommunications equipment should always be of the type which include an isolated ground. Ground fault in ordinary electrical service is a problem, as is interference from other equipment connected to your electrical system. Your buildings management, or your electrician can give you more information about the concerns having to do with adding service for this purpose.

All electrical telecommunications equipment should be grounded, and should be connected to surge protection devices, as well as battery backup.

Whether or not local codes reflect this, phone systems are in fact life safety devices. In the event of a building mishap, such as a power outage or a fire which could cause discontinuity of service, there should be a means of operating your phones via battery.

Relying upon individuals who may have cellular service is at best, a spotty contingency plan, and is not recommended. Battery backup also adds another layer of protection to your delicate phone switching equipment.

The day-to-day advantage of battery backup is that, should the power fail, you will not lose "memory based" programming, such as pre-programmed buttons or speed dial. In the normal course, these features are lost when the phone switch is powered off.

VENTILATION AND ELECTRICAL PROTECTION

Telecommunications closets should be in locations which can by means mechanical, or natural remain ventilated and reasonably cool. A good point to shoot for, realistically, is between 68° and 70° Fahrenheit. In an imperfect reality, many equipment rooms will venture into the 73-75° range, and may peak at higher temperatures still, depending on your region and weather. Keep in mind, however, that your equipment should never be allowed to operate at a consistently higher temperatures than 75°, because it will significantly shorten their operative lifespan.

Bear in mind also the warranty of your equipment. Some electronic equipment is present in all telecommunications closets. Most warranties that are included with telecommunications equipment say something about abuse. Abuse includes inadequate ventilation and cooling.

GROUNDING AND BONDING

The National Electrical code has an entire chapter about grounding and bonding. Grounding is very important for any electrical or electronic device. This book will not discuss particular grounding methods, as they vary from installation to installation, and evolve over time.

However, it should be said that every piece of equipment, every wire which runs outside, and every equipment rack should be grounded to a post which is visibly in the ground. If you are on an upper level, there should be a centralized ground wire running in the building for your telecommunications equipment. If you can't find it, check with your building management to find out where it is.

Telecommunications equipment should not be grounded to the same conductor (ground wire) which is used to ground the electrical equipment. There is specific hardware and wire for grounding telecom equipment. This can also be explained in greater depth to you by an interconnect system provider.

As an end user or administrator, it is only important to realize that WITHOUT GROUNDING, YOUR PHONE SYSTEM WARRANTY MAY BE INVALIDATED.

Something to keep an eye on is the labeling of grounding wire. Like all communications wire, according to ANSI/EIA/TIA section 606-A, all grounding wire must be labeled and numbered to identify it in a plan of the system room, and to differentiate it from other pieces of grounding wire. All of this information should be recorded in a paper drawing or some such plan, which is created by the wiring installer.

PHYSICAL PROTECTION

Protect your organization's telephone equipment. Put out the word at your organization that your phones, as cool and funky as they look, will not work at the homes of your staff or volunteers. Telephone switches don't get stolen often, but telephone stations do get "borrowed" from time to time.

As we will discuss later, it is a good idea to put into the instruction sheet for your phone system topics having to do with (1) not trying to use foreign equipment with your phone system, and (2) not trying to use your phone at home. If you are finding problems with one of these (the former is much more likely than the latter), this communication might be beneficial.

PATHWAYS AND RACEWAYS

Pathways and raceways are defined in great detail in the National Electrical Code, sections 770 and 110.6,. A pathway or raceway is the application to which wire is able to get from the jack in the wall in the office, up the wall, through the ceiling, down the stairs, and into the wire closet.

There are as many different applications to hold, route, and manage wire as there are types of buildings. Your organization may need certain types of special hardware, such as surface mount, to get phones to key locations.

Some Perspectives on Raceways:

There are different priorities The building owner's desire is to have invisible wire. Owners always want to hide the wire, so as to maintain the aesthetic beauty of the space.

On the other hand, the installer's rule of thumb is to pull the wire in the easiest way possible, the cheapest way possible, which requires the least amount of time and effort, while still maintaining the code requirements.

The service personnel cares neither about ease of installation or aesthetics. He wants the wire installed in the most visible, easy-to-trace and access way possible, preferably low to the floor, so that he can get to it without having to take the ladder off of his truck, and venture into the ceiling.

I worked as a system manager for an organization whose offices were located in a house which was built in the 19th century. The walls of the house were made of stone, and the interior walls were plaster. It was necessary to put new telecommunications wire into the organization. We were told by the owner that there could be no surface-mount raceway, which was a product which is visible after installation. This made the installation very time consuming and very difficult. Left to the devices of the installer or the vendor, we would have ended up with a surface-mount product, but the ownership insisted, and won the argument.

By the way, there is nothing wrong with surface mount raceway. There are many attractive applications which sell under various brands like Wiremold, Panduit, and others. Sometimes, there is no practical way to get wire from point A to and to hide the wire without tearing apart the walls and ceilings to do so. Comparative to hiding the wire, this is generally not as attractive an option, but there are instances where the installer would do more damage to the building without using the surface product.

What can you do as and end-user? Make sure that this type of issue is specified and understood in the installation contract prior to getting involved with a vendor. This is one of those "workmanlike" issues which becomes cloudy when the problem arises.

A tip is to ask the vendor to look at your installation, and to indicate whether they have done this kind of work elsewhere. Inquire as to the success level, and talk to others who have used your prospect to get an idea of their level of quality and ability to handle your job.

FIRE STOPPING AND LIFE SAFETY

There is a basic rule of thumb with fire safety: There is a presumption that buildings are built to the letter of the local fire codes, to isolate smoke and fire between "zones", such as floors or between hallway sections. Local, and national fire codes specify many aspects of building construction to predicate certain levels of fire-retardant property. When buildings are built, they are inspected by fire marshals, who look to see that the buildings have been built correctly. Unless the fire marshal misses something, or gets paid off, the building should technically be able to resist the spread of fire in accordance with the local code requirements.

In reality, even honest marshals miss things, and no building is perfect. There are holes which nobody can see, and things are missed. How does this affect you? It's simple. Don't let your Communications system cause someone to die by defeating the level of fire safety in your building.

Fire safety in a building is like water in your basement. There are few basements (in the Northeast, anyway) which do not have problems with either moisture, or water seepage. A few holes in the system are not a big deal, but many holes cause a flood every time it rains. Just like your basement, propping open a fire door, cutting an unseen hole in the fire-barrier wall above the ceiling can cause trouble. Maybe not on it's own, but an aggregate of these problems can cause real trouble, and may cost someone's life. That's why the NFPA takes this stuff so seriously.

Remember that thing called "workmanlike fashion"? Ever wonder how things get sloppy? New buildings usually aren't sloppy. The problem with many things is that somewhere along the line, someone did something small which was sloppy. Perhaps they cut a small hole in the wall above the drop ceiling, pulled their wire, and walked away. Maybe they didn't know about pulling wire, or they didn't care about it. Maybe they were a volunteer, or someone's boyfriend. Maybe they were just starting out, and weren't properly supervised.

The next guy comes along, sees the mishap, and does what comes natural to some: he follows suit. Now there are two holes in the ceiling, unseen.

Well, eight unseen holes later, this building has a fire, which is noticeably thwarted by the fact that there is a serious gap in the wall, out of view.

Reactionary? Perhaps, but the cost and effort to finish the job is usually so small as to be laughable. Many good products are sold to prevent your breach of the wall to be a problem should disaster strike. These fire-stopping products are inexpensive, and are readily available. It is for this reason that I am stupefied that so few structured cabling contractors use them.

Making things right is the real advocacy here, not necessarily limited to fire safety, but in terms of neatness. Don't give the next contractor down the line any excuse to do something in a half-assed way.

What do you need to know as the system manager? Not too much, but make sure that the company contracted to install the wire knows that you know about fire-stopping. 3M, and HILTI, two companies who make a lot of construction-industry products, make sealant which is fire retardant and is specifically designed for holes made when installing wire.

PLENUM AND RISER CABLE

Two types of cable which you should be cognizant of are Plenum cable and Riser cable. Table 800.50 in the 2002 NEC defines these two types of cable which are important, called MPP (Multipurpose Plenum rated), and MPR (Multipurpose Riser rated) cable.

The key ideas:

* Plenum is for above the acoustic tile ceiling because it burns the least noxiously—it creates the least amount of noxious, black smoke when it burns.
* Riser is for one floor to another, because it burns at a low vertical rate, less than P.V.C (low-cost communications cable).
* Plenum is better than riser, in that it burns less noxiously and at *the same rate* as riser cable.
* Riser is better than P.V.C., which is the worst and should not be used anywhere.
* P.V.C is the cheapest, Plenum is the most expensive.

Plenum cable is communications-grade cable which is installed in the ceiling. It has a different kind of plastic coated shield, and is called "plenum" because it can be placed there safely. Plenum cable will not produce as much of the toxic fume if a fire starts in the area above the acoustic tile ceiling.

Because most newer buildings use the area above the tile ceiling as an intake for air, fumes in this area could be more deadly than either smoke or fire. This wire will not contribute to those fumes, as is the case with normal, non-plenum wire.

Riser cable is safe to use to run between floors. It's plastic casing will not burn vertically as fast as non-riser type cable. This prevents a fire from being able to spread as quickly between the floors in the building.

WIRE TYPE AND CATEGORY CLASSIFICATION

Although telecommunications is changing every day, for the purposes of today's technology, and for the purposes of this book, we are assuming that 4-pair copper wire is being used for the horizontal (in-ceiling) and vertical (between floor) cabling systems.

There are several types of wire available. There is the less common shielded wire (which has a thin foil shielding for the pairs inside of the plastic protective casing). This wire is called Shielded Twisted Pair Cable (ScTP). Ther is the more common UTP cable, (Unshielded twisted pair) which has no foil shield. Both of these types of wire look physically the same, and have the same four pairs (8 wires) made of 24-gauge copper wires inside of the plastic casing.

For these two types of wire, There are several categories of grade. Categories indicate the degree of twist in the wire, and act as an indicator to the level of quality of the cable.

There are several categories of copper wire, numbered 3, 4, 5, 5e, and 6. These numbers are vastly more critical in high-speed networks, such as data communications systems. Their determinant factor is the amount of speed at which signals may be carried through the wire with an acceptable amount of signal distortion, and loss due to interference and noise.

CATEGORIES

* LEVEL 1: Untwisted station cable, and old phone wire. Acceptable for voice application only, and may not work with newer phone systems
* LEVEL 2: 1 MHz rated voice-grade wire, such that is installed in residential construction from the 70's or 80's.
* CATEGORY 3: 16 MHz rated voice-grade wire. Commonly used in the past for phone system installations and older computer network installations of less than 10 megabit speeds.
* CATEGORY 4: 20 MHz rated computer/communications grade. Not commonly seen, as it is no improvement over Category 3.
* CATEGORY 5: 100 MHz computer/communications grade. Very common in newer construction including computers and phone installations. Minimum grade currently recommended by BICSI for new installations.
There are two other grades, called 5e, and 6, which are very important for the new computer network installations. These groups will not be discussed here, because they have no bearing on current technologies in basic voice-only telecommunications systems.

As of this writing, BICSI says that for new installations, no less than Category 3 certified wire should be used. My opinion is that no less than Category 5 wire should be considered, perhaps even 5e or 6, if you are planning, as most people do, to have a computer network. As all of these quality grades contain the same color coded 4-pairs of wire, the physical link will work with any of the three. The examples in the book assume either Category 3 or higher of quality grade.

The only other crucial information about wire is in the type of conductor used. There are two types of wire, those using strands of much smaller copper wire which are wrapped by machine together to make the equivalent of 24-gauge wire. There is also a different type, called "solid core" which is one piece of 24-gauge copper wire. Solid core cable should be used in infrastructure wiring (between the frame and the station). Stranded wire, because it holds up better to physically being twisted and pulled, should be used for the short "station cable" run between the jack and the telephone base.

ATTENUATION CROSS-TALK, LOSS, AND NEXT (NEAR-END CROSS TALK)

When deciding the location of the wire closet, it is important to calculate the distance between the wire closet and the furthest telephone station. Different manufactures use different voltages, and different technologies to drive their phones, electrically, speaking. Check with the manufacturer of your phone system to determine the maximum length between the phone switch and the station.

Copper wiring produced today is plagued by two problems, the first is a factor of temperature, and distance, and is called *attenuation*, or loss.

Structured cable installers and testers look for ACR, which stands for Attenuation to Cross-talk Ratio. Attenuation is the reduction in signal strength over the length of the cable, and frequency range, the cross-talk, as we discussed, is the external noise that is introduced into the cable. If either of the two increase beyond an acceptable range, the data signal will be garbled, and unrecognizable.

ACR is the most important test, because it represents the total overall performance of the entire link.

Attenuation is a form of loss. Loss is the amount of signal lost at the far end of the wire, versus the original signal strength at the begining of the wire. This is no different than the kind of loss inherent to sound or radio waves. This is why you can't hear people talking across the street, or why distant FM radio stations won't come in clearly.

Cross-talk occurs because in 4-pair wire, there are 8 wires, each operates as a pair. Multi-pair cable is said to be "balanced", meaning that in each pair, one wire is carrying a voltage of +5 volts, and the other of—5 volts. If the voltage in the wires is balanced, the electromagnetic fields which are generated by the voltage in the wire cancel each other out. This cancellation is a factor of twisting each pair of wires together. The tighter the twist in the wire, the more effective the cancellation of noise, and the lower the amount of "crossed signal" between the pairs of wire.

Electromagnetic interference, of the same kind as cross-talk, is also generated by outside sources, like electrical supply equipment, and motors. These pieces of equipment should be kept at a distance from your phone equipment, so as to minimize this type of interference.

Telephone equipment, whether digital or analog in architecture, is subject to operate within a certain range of electrical voltage and with a certain amount of voltage differentiation. The effect of both cross-talk and loss is a blur in the differentiation. The nature of copper data communications wiring is such that over a long distance (100 meters or so), the power of the signal will weaken significantly, unless artificially amplified through some means. Problems having to do with voltage drop range from minor (a station "cuts out" while in use or during peak activities, such as ringing), to major (unexplained equipment failure).

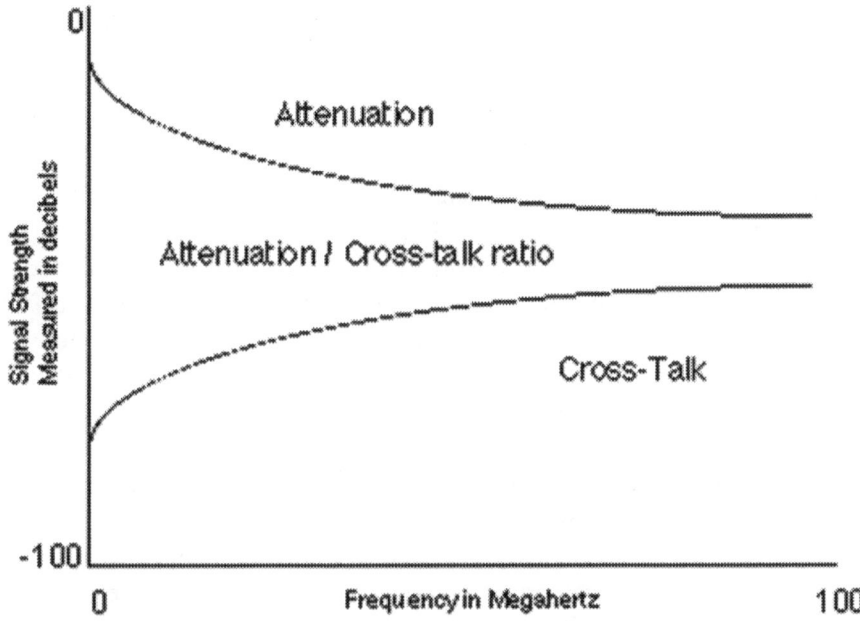

Figure 1-12: ACR effect as a factor of distance in copper wire. Measured at a fixed ambient temperature. If the range of Attenuation meets the range of Cross-talk, data will not be able to be transmitted over the cable in a recognizable fashion.

Manufacturers indicate that loss is inhering to the physical properties of copper cable, because there are always slight imperfections in the copper itself.

Because phone system manufacturers are aware of, and expect this kind of problem to occur, they design their telephone systems to operate using (1) certain quality and manufacturing grade of wire, and (2) maximum distance from frame-to-station. System manufacturers always specify these requirements and distances in their product literature.

Wire installation certification has a great deal to do with preventing an excess of these two inherent weaknesses of copper wire. The amount of twists over a distance in the wire prevents the cross-talk from overtaking the signal to a degree where the equipment will no longer be able to function. The amount of length, no more than 90 meters for Category 5, determines the amount of distance before attenuation loss begins to affect the signal to a degree where it falls out of the range of the standard.

One manufacturer, Telrad, specifically designs their systems to be able to handle a mix of "short" and "long" runs to telephone sets. The station cards can be set to handle a certain number of stations (8 or 16 per add-in card, depending on the type of card). When the card is set to handle "long runs", the maximum distance increases significantly, but the number of telephone stations which can be handled by the card halves (e.g. a 16-station card can now handle only 8 stations).

So you think that you have exceeded the manufactures length but the phone still works? If you exceed the maximum length, it does not necessarily mean that the phone will not function. I have worked with many systems, especially those which are older, which will work on much greater distances than specified. Why is this? One reason, perhaps the systems were intended to work with untwisted wire. Many older systems will experience much greater success with distance when twisted Category 5-type wire than with the older Category 3, or station (Untwisted) wire.

However, it is not recommended to do this, despite the fact that it seems to "work". Because voltage drops over the length of the cable, the amount of voltage at the end of the cable with the telephone station will be lower than the end at the telephone switch (the power origin). Longer cable has higher levels of drop, which means much lower voltage at the end of the wire.

Significant drop, especially where length approaches or exceeds the recommended threshold for the phone will likely do damage to the sensitive electronics within a digital telephone. These phones actually can have greater damage done from undercurrent than from over-current. One thing about PBX telephone systems is that the stations must always be receiving some level of electrical power from the origin. Extending the distance between the phone and the switch will make the power supply have to work harder, this in turn generates more heat, which unchecked can damage internal parts of the phone system, most notably, the power supply itself.

For administrators and end users of such systems, checking with the interconnect vendor is a good idea. Look for locations which are excessively far away from the phone equipment room, such as in converted attics of houses.

Additionally, make certain that when measuring distances, that the real distance of the wire is measured, including walls, through ceilings. If the wire must take a circuitous route to a far-away telephone, this can make the difference.

WIRE TERMINATION

There are no less than a dozen major manufacturers who make products in the coveted market of "frame to station" hardware. These products include wall-jacks of a near infinite variety of configuration and color, trim plates and installation hardware, wire ties to hold wire, and panels, racks, and blocks to put into your telephone room.

The rule of thumb is to stay with the same brand of wall jack in an installation. This facilitates repair and installation, and allows you to stockpile a certain amount of extra parts for your system.

Additionally, I try to stay with standardized, commercial parts of the same brand. There are a few manufacturers selling "systems" for structured cable (phone, network and video wiring packages). These systems, I believe, to be overly expensive, and not particularly more useful than using ordinary, commercial-grade hardware.

The problem with these integrated "solutions", which are usually aimed squarely at the hapless homeowner, lies in the fact that they are generally proprietary to manufacturer and model. The jacks won't work with other brands, or look strange because of size or color. The equipment won't fit in the rack of a different brand, because the racks aren't standardized.

Other than these oddballs, most major electrical application suppliers, like Pass and Seymour, Leviton, and Hubbell, make fine data and telecommunications equipment, which is of high quality and reliability.

Aside from the brand of wiring products, there are generally three types of wiring standards when dealing with telecommunications equipment. These standards are defined by EIA/TIA, and are referred to as 568A, 568B, and USOC, or Uniform Service Order Code.

These standards indicate the order for which wire is lined up both in the patch panel, the punch-down block, and the wall jacks. Following these standards ensures that no "crossed wires" occur in your house cabling. All of these standards imply 8-wire (4-pair) copper cable.

Both 568A, and USOC are older standards, but is still used with some phone systems. Much more versatile is the 568B standard, which is used almost universally in new installations.

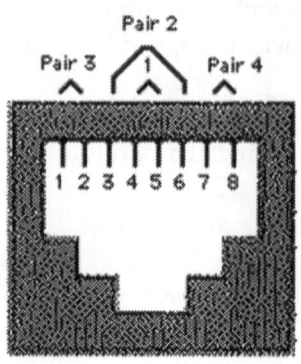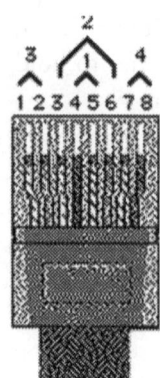

TIA 568 A STANDARD

PAIR 1 BLUE
PAIR 2 ORANGE
PAIR 3 GREEN
PAIR 4 BROWN

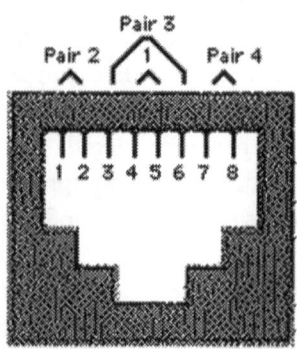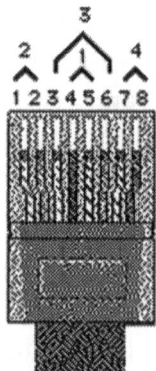

TIA 568 B STANDARD

PAIR 1 BLUE
PAIR 2 ORANGE
PAIR 3 GREEN
PAIR 4 BROWN

Figure 1-13: EIA/TIA 568A, B Standards for wire mapping

LABELING OF JACKS, PANELS

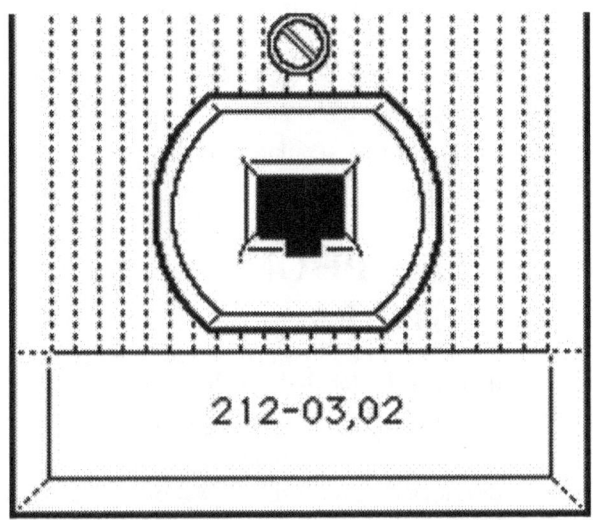

Figure 1-14: Jack label for Room 212, Jacks numbered 02, and 03

ANSI/EIA/TIA standard 606-A dictates that wall-jacks and panels need to be labeled using a standard schema. There are many different ways to number jacks, and for a small system, you may wish to use the office or room numbers for simplicity.

There should be a label on both the jack on the wall, and on the corresponding cable, and port on the panel in the equipment room. This greatly facilitates moves, changes to the system, and the periodic service which will occur to your telephone equipment.

There are several companies who deal in automated software which generates, using custom printer forms, pages of labels for jacks and panels. These forms can be invaluable, and can make an installation look much more professional than the ubiquitous laminated label maker.

The key, though, is legibility, organization, and durability. Make certain that whatever scheme you use is reasonably simple, and continues to be valid even if people move.

Labeling jacks "Mary's office" or "Secretary" is not advisable, as in years to come offices may move and faces will almost certainly change.

Most of the time, facility maps designate room numbers, and numbers are often posted at entryways. Using this numbering scheme is straightforward, and will help service people to easily locate wiring in the offices.

Additionally, the number of the jack should be handwritten or labeled using standard numbering labels (available at any electrical supply store) on the wire itself, about a foot or two back, so that it cannot be spliced away. This makes the system impervious to changes in the faceplates on the wall, and painters who reap havoc with jack labeling.

LABELING OF TELEPHONE SETS, DESIGNATION STRIPS

Just as the jacks and panels can be labeled, so should be the telephones, number ID's (which identify the station extension and telephone number) and the buttons.

These paper strips are referred to as "DESI'S", short for designation strips.

Designation strips are paper forms that come perforated, or cut, to fit into the telephone's regions where there are programmable buttons or where the phone number and extension information are displayed (usually on or under the handset).

Because designation strips are, by nature, proprietary to the model and brand of telephone, they are usually only available through the manufacturer or vendor of the particular phone system which you may have.

The telephone will come with one, or one and an extra, designation strip.

Number these in a careful way, using an ink pen or preferably a typewriter. Use extension numbering, rather than people's names, and plan where buttons are to be located on the phones for Redial, Hold, et cetera, to minimize the need for replacement.

What if I need replacement strips? Major branded telephone systems will have the strip available for years to come. There is an emerging alternative to this, however, which may be more attractive to you. There is at least one company who manufactures replacement strips for a vast number of different telephone systems. These replacement strips come on a printer form, which along with a software program, can be run through a computer printer.

The printer form then can be detached, so that the pre-printed designation strip need only be inserted into the telephone.

This allows a simple, neat, and professional alternative to white-out, and makeshift replacement strips.

BATTERY BACKUP SYSTEMS

Battery backup is an oft forgotten necessity with regard to a telephone system. All PBXs and key systems operate, at least in part, relying on AC line power. What happens if the power cuts out? The phones go dark and nobody can make any calls.

At out installation, we installed a battery backup which will run the voicemail and phone switch for 48 hours. This amounts to a battery-backup system which requires a bookcase sized shelf of 48" by 70" to hold the batteries and the battery-backup system. Our decision to install such a powerful backup system had to do with the fact that we have people living on our premises. Being a Catholic church, we have full time residential clients, who may have need of the phones at any time, night or day.

More than any other service in your office, or ours, telephones are a life-safety issue. The inability to be consistently able to make a phone call could cause someone a lot of difficulty, or potentially physical harm. We did not want to entertain this possibility, so we installed a large battery backup system for our telephones.

Additionally, we installed into each phone closet an ordinary telephone. This allows, in an emergency, an individual to pick up the phone and call 9-1-1, or the electric company, or anyone, provided that the phone network itself is working.

Despite the increase in the commonality of the cellular phone, there is still a necessity to have, and a place for, regular phones. Cell phones, as shown in the 9/11 tragedy, are not impervious to system-wide problems, and outages which were not felt by local exchanges and normal telephones.

It is advisable to have both surge protection, and battery backup on your system which will allow at least 8 hours of backup power. To figure out what you may need in terms of power requirements, talk to your system's vendor about the power requirements for your phone system.

Section 2:
Features,
Features,
Features

BASIC FEATURES OF ANY PBX SYSTEM

There are basic features inherent to any business telephone system. These features include Speed (Memory) dialing, Hold, and Last Number Redial. Some of these features are very straightforward, and require little explanation. Others are inherently more complex, and are not so familiar, unless one has previous experience with business telecommunications systems.

CALL TRANSFERS

The ability to transfer a call is a marked feature that separates a residential phone system from a business phone system. There should be a feature to signal another station, talk to the person at the end of the station (without the outside caller hearing). Subsequently, there should be the ability to have that person pick up the line with the outside caller, or have that call transferred to that extension.

When transferring outside callers, there is a school of thought that says that all transfers should be confirmed. Just as a person who entered your facility looking for help would be assisted, your callers should also be assisted. Except for the annoying army of salesmen, don't send them to voicemail hell, or transfer them to an extension which may be unmanned.

Verify that someone is on the other end of the transfer, and perhaps tell the individual how to reach that person, should the call fail to transfer correctly. Some organizations will not wish to give out this information to all callers, which can be understandable.

PROGRAMMABILITY

PBX systems range in the level to which they can be customized. This level of customization is referred to as "programmability". Some basic systems, such as the AT&T Merlin classic line, offer very little in the way of programmability. Station extensions are fixed in the hardware of the switch, and cannot be altered.

Additionally, there is no need for a centralized "mother phone" to program extension buttons or other features.

On the other side of the scale are full-fledged computer programs which create a "memory image" which is stored on the disk, and is transferred to the memory of the phone switch via a serial cable or computer network link.

Basic telephone systems can be installed, turned on and learned, with very little configuration or maintenance necessary. More complicated systems may require an evening or two with the manuals, and a few meetings with those on the staff about how features should be set up within the system.

CRT TERMINALS, LAPTOPS, AND OVERLAYS

The computer program used to configure the Telrad digital phone systems, for instance, requires a CRT terminal running DOS.

The program has different screens, which create an "image" in a file on the disk of the computer. At the push of a function key, the disk image is transferred to the memory on the processor board of the phone switch via an RS-232 serial connection.

In many ways, this is an excellent solution. For the system that I manage, the disk image is easily backed up to floppy disk, such that the entire configuration of the telephone system is retained "offline". This includes, for a 100-station system, all of the memory buttons for the entire system. You can imagine the trouble with re-configuring all of the speed dial, and options for 100 stations!

Off-site, a copy of this image resides at the company who services our telephone system. This way, even if catastrophe struck, a backup of the system would exist and could easily be replaced.

For many customers, the service person would carry a laptop or portable computer to their site, and would be able to see and change the configuration of the system easily. This negates the need to have a dedicated computer in the equipment closet. For customers with only casual necessity to interact with the configuration of their system, this may be a desirable option.

Alternatively, smaller systems exist with no such ability to program using a CRT terminal. These types of systems generally use a "mother phone" arrangement. The mother phone may be a larger phone with more buttons, and may have a simple LCD display.

The display, and an overlay is used to program the phone system from one location. An overlay is simply a piece of paper with cutouts for the buttons. The overlay is placed over the normal insert, and the programmer can see what each button configures. Normally, there is a switch on the mother phone which allows the user to move into the programming mode versus the ordinary mode.

AT&T's Merlin, and Partner-style phones have a three-mode switch, which is very handy. There is an ordinary use mode, a "test mode" which signals a beeping noise, and flashes every dual-mode light on the phone (for testing). There is also a third "programming mode" which is used to program the functions of the memory buttons.

In more advanced systems, such as AT&T's Partner line, all functions for all extensions can be programmed via a single phone using the overlay and numeric codes.

HOLD, MUSIC-ON-HOLD

Hold allows one to pause the call and hold the line. The effect is that the call is held, but one does not lose the call. This feature is very useful if a subsequent call must be made, or if some other interruption occurs.

Hold is a very basic feature, and does not require that a phone system be installed. However, a multi-line phone, and more than one phone line is required to be able to make another call while holding a call.

Hold is a very dangerous feature. Don't put your callers on hold for extended periods of time. If possible, instruct your operator to ask the individual whether they would like to hold, and give them an idea of how long, or whether they would like to have their call returned.

A good organization will have some kind of follow-up, if by no-one else, the operator. Ask your operator to follow-up with the person at the next stage, to make certain that a call-back did occur in a reasonable amount of time.

Should someone need to be kept on hold for a long time, check back with them often. Re-open the line periodically to make certain that they know that they are not forgotten, and to re-assure that the wait won't be too much longer.

CUSTOMIZED RINGING

In an office with a lot of desks, and extensions, it can be helpful to customize the ring of each station, so that people can begin to identify the tone of their phone versus other phones. This feature is excellent if one needs to be able to leave their workstation, but still recognize that their phone is ringing.

This is a feature often associated with business, versus residential telephone systems. Most simple business telephone systems allow that ringing on individual stations can be turned on or off, delayed, or changed in tone to audibly identify a particular station.

SPEAKERPHONE, HANDS-FREE-INTERFACE

Speakerphone has been around for a generation of telephone users. Early speakerphone was an external attachment, with a large speaker/microphone housed in an enclosure. It was a feature for business telephones, and was available through the phone company.

Newer technologies abound, and good quality speakerphone is an option on any modern telephone system—either as a standard, or optional accessory.

Speakerphone may well be the reason why you are interested in a more complex telephone system. Hands-free is a great help, especially if you deal with clients, or need to conduct business with a third party who is on the phone.

With many telephone systems, the speakerphone functionality is a feature of certain models of telephone stations. Bear this in mind when you figure out what kind of telephones you need to buy.

One company, Polycom, is producing an analog telephone product (not a phone-system phone) which is electronically designed to be used by a group of individuals as a conferencing device. The phone, which looks like a black plastic triangle, has speakers and microphones and is designed to sit on a conference table or a desk. It is set up to be able to receive audio from the three points of the triangle. The circular center is a speaker grille, which covers a sizeable speaker designed to allow the other party to be heard clearly. This unit has external amplification which, unlike many telephones, requires that it be connected to an electrical source as well as to the telephone network.

Figure 2-1: An example of a conferencing telephone

A word on connectivity: There are two classifications of specialized telephone equipment. There are those devices which are designed to work with individual manufacturer's phone systems. These devices can be portable phones, door phones, and the like. They are usually manufactured by either the phone system manufacturer itself, or a subcontractor to that manufacturer for use with a particular line of product. Conversely, in today's marketplace, there are devices which are designed to be universally compatible with the ordinary telephone network. These are usually referred to as "analog" devices. Analog telephone equipment, like answering machines, modems, and the phone pictured above, are abundant. This equipment, however, can be a problem when dealing with a fully integrated telephone system.

The first problem with this equipment is that it cannot directly interact with your other phone stations. It must operate in parallel to those systems.

For instance, if you have a phone system, and require automated message service, your least expensive and most readily available option is to install a telephone answering machine. As we will see, there is a world of difference between basic phone answering machines and phone-system answering systems, which are, by nature, integrated into the operation of the phone system.

Similarly, facsimile machines, and specialized equipment like door phones, and all-weather phones, may only be available to you in a generic sense, not specifically for your phone system.

Telephone system manufacturers will sometimes offer a bridging device for this kind of equipment. AT&T, for instance, offered a device which could attach, by means of an external adapter, a modem or facsimile device to a telephone station within its Merlin telephone system. This allowed one to have a fax machine, or modem at any location where a Merlin telephone was installed.

Other telephone systems, such as the Telrad system, allow you to purchase "analog station cards", which allow normal telephone equipment to be directly installed to the telephone switch without having to connect the device to its own telephone line. These cards allow you to install a portable or conference phone at an "extension", albeit obviously without the advanced feature set inherent to the digital telephone.

Many people will opt to use this functionality, as certain types of devices, such as portable phones, are either not available to small systems, or are prohibitively expensive.

DO-NOT-DISTURB

Do-Not-Disturb, often called "DND" for short, is a very useful feature. With the simple press of a button, your phone station in your office can be set to either receive calls (DND is off) or not receive calls (DND is on). If your organization has a system with voicemail, DND can be programmed to put the caller directly into the voicemail box without ringing the telephone, and thus disturbing you.

This can be very useful if you are having a meeting, or otherwise do not wish to receive telephone calls.

PRIVACY AND CONFERENCE CALLING

One facet of a very small phone system is the method to which lines are selected. Some systems allow the user to depress a button, while on the phone, which prevents others using the system to inadvertently pick up the in-use line and "join" the call.

Conversely, most all business phone systems allow the "conferencing" of stations around one inbound telephone line. This allows persons to join the call who may wish to become part of the conversation.

RUDIMENTARY MULTI-LINE CONNECTIVITY

Telephone systems are marketed and engineered at a very basic level on two factors: The number of stations and the number of lines. There are no modern systems which allow fewer than four inbound lines. This means that four telephone numbers can be attached to a telephone switch. Depending on the configuration and complexity of your phone system, you can make stations pick up the next available line in the list, allow particular stations to use particular lines, or disallow certain stations from making outbound telephone calls entirely.

This selectable line feature is very useful for organizations who wish to make rudimentary phone service available to clients.

I have one such organization whose mission it is to serve the homeless population of an urban area. Part of their mission is providing a place where the phone service is free to the clients, and is available so that they may be able to apply for jobs, or get some kind of assistance. So much of successfully getting assistance is being able to maintain the contact that reliable telephone access to these individuals is key to the success of the program.

The phones, however, have been set to dial out only on lines which have no long-distance service. Because of the obvious expense and problems associated with allowing free long distance, it is important that these phones only use certain lines, and only ring when certain lines are called, so that the individual can receive calls at the location consistently.

Other phones at the installation, however, are set up to dial out on lines which have long distance service, so that client assistants can make long distance calls to other facilities and assistance networks.

The programming of these features is done through the phone system itself, and is not reliant on the telephone network or the local phone company.

MEMORY BUTTONS

A great feature to have on a phone system is the ability to have speed dial, both for your own contacts, and a global speed dial for groups which are called by all of the members of your organization. Memory buttons are standard on virtually all phone systems, and are easily programmed by the user of the telephone.

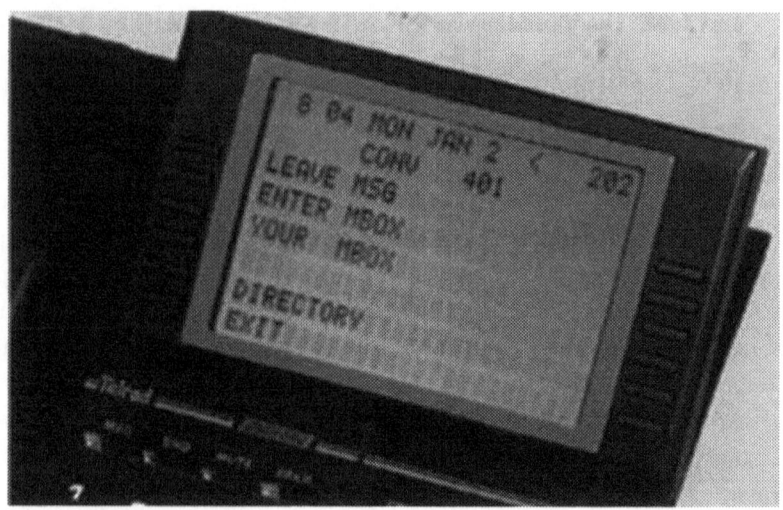

Figure 2-2: Example of a digital display on a des telephone

LCD SCREENS

As the state-of-the-art in digital technology progresses, there is an increasing desire to allow visual prompts to be added to telephone systems. An example of the visual interfaces which I have seen on a small system is the one shown above. Manufactured by Telrad, this screen allows the user to move through prompts normally only heard through the automated attendant. The prompts are visually displayed as one would see on an ATM machine. Note the unlabeled buttons on the sides of the display, which allow the buttons to change as the contextual situation displayed on the screen changes.

Certain items may be displayed on the phone's screen:

TIME-DAY-DATE

The Time of day, the day of the week, the date as specified by the internal clock/calendar of the phone system.

CALL TIMER

A visual "Stopwatch". This is an indication in hours/minutes/seconds of the elapsed time that you have been engaged on a call

INTERNAL CALLER ID

The extension number, and sometimes name of the phone which you are calling. Some newer systems allow you to name extensions within the phone switch, so that you see "Mary" when you call the phone in Mary's office, rather than "Extension 1003". This can be of great help when Mary calls you, because you can glance down at the screen and see that she is calling you, rather than Ed.

CALLER ID

Ordinary caller ID can be integrated with the appropriate services, and usually a "Caller ID Adapter" which is installed to the telephone system in the equipment closet.

A WORD ON CALLER-ID:

Perhaps more so in the world of the not-for-profit than in other venues, there is the danger of irate, disturbed, and even demented individuals calling your facility. Some of these people are regular callers, who are known to the organization. Others, of course, will be unknown.

For this reason, it may be a good idea to have caller-id services. Caller-ID allows you to at least have a starting point for dealing with the recurring abusive caller.

The irate caller may not be abusive, but should be handled gently, and transferred to a patient person who may be able to legitimately hear out the caller, and perhaps even solve a problem within the organization. At the very least, this may be a method of diffusing a situation which cannot usually be effectively handled by the operator.

Abusive callers, and those who are disturbed, should be handled gently, but firmly. At the point where the person becomes abusive (for no apparent reason), the call has ended. Explain to the caller that he or she should not call again, until they can handle themselves appropriately on the phone.

ADVANCED FEATURES

Beyond the basic features inherent to most any business phone system is another subset of features found in slightly larger, more advanced systems. Some of these features are basic add-ons, while others are architectural differences between simpler and more complicated phone systems.

EXTENSION DIRECT DIALING AND INTERCOM

One basic architectural choice need be made in selecting a telephone system is the level of extension connectivity required.

Some basic systems, like the older classic Merlin by AT&T, can only interact with stations using the "Intercom" feature. Pressing the Intercom button drops the user to a different sounding dial tone which is generated by the phone switch itself. The user dials the extension (in the case of Merlin Classic, the extensions are fixed and are 1 or 2 digit numbers). The user dials the number, and the destination extension beeps to indicate that the intercom has been initiated. The user can then speak to the destination through either the speakerphone or the handset application.

Larger systems can actually dial internal extensions by dialing their three or four digit number. This allows the user to use their internal extension as you would normally do to make an outbound phone call. Other phones on other desks will ring, and the call can dealt with as would an outbound call.

Internal extensions in medium-sized to very large systems can be directly dialed from outside the system. These calls would ring on the internal extension, without having to be transferred by an operator. If voicemail was present, the call could be handled automatically if the person was not at their phone station.

The size and layout of your organization will determine your level of need in terms of extension direct dialing functionality.

SPECIALIZED STATIONS, INDUSTRIAL APPLICATIONS, ALL WEATHER APPLICATIONS

Phone systems are usually designed to be positioned to certain kinds of customers. Telrad, for instance, is a great system for inter-departmental use, because multiple secretarial multi-button stations can be handled, and calls can be routed between a variety of departments easily by the automated menu system.

However, Telrad does not have a great deal of outside or industrial grade user equipment. If you need to have a metal phone for a harsh environment, or a waterproof phone for outside, many manufacturers will not be able to supply these kinds of specialized things. Because only Telrad phones will work with Telrad switches, and this is the case with almost all phone systems, there is no ready option for an outside phone, short of putting a regular inside phone into a waterproof box.

Looking into the options should command a certain amount of importance when auditioning telephone systems. Especially if your organization has out-of-doors phone needs, you should ask if these options are available.

Some systems also offer options to connect the phone system to security systems, and locks for doors. This is invaluable if your organization, as many have had to do, needs to lock the outside door during the day. The receptionist has the ability, with some systems, open the door lock from the phone station at her desk.

LARGER SYSTEMS

Larger systems are usually a combination of multiple system cabinets and more extensions, and more wire. Some systems will never grow in this way, because of the nature of the size of the organization. There are only so many phones which will be necessary to operate your organization effectively.

This book really isn't about large-scale phone systems, but there should be the provision to add about another 1/3 of the number of phones which are to be installed. Plan for expansion, especially, if your organization needs to live with the system for more than five years. Many phone systems cannot handle expansion, and thus are useless beyond the small boundaries which the manufacturers have set forth.

If expansion is important to you, look for systems which can be linked together to form larger systems. Look for systems which can be expanded through the use of stacking cabinets, or expansion cabinets attached to the main system.

LINKING CABINETS

Expansion of a phone system usually means that another cabinet will be added. Adding another cabinet allows one to add another power supply, and to add modules or cards to the system. Cards and modules are the means by which additional phone stations can be attached to the telephone switch. Advanced options such as auto-attendant (explained later) and voicemail also attach to the system via cards or modules.

When physical bounds exist in the cabinet architecture, the only add-on options exist through supplemental cabinets. If no such option exists, be aware that you may "hit the wall" so to speak, and not be able to add on additional extensions to the system.

Some systems were designed to be able to co-exist or integrate with additional systems. This allows a great deal of expansion, in that the systems can be linked by purchasing another phone switch and more extensions.

TRUNK GROUPS AND OUTBOUND HUNTING

Phone lines which are connected to your system are referred to as either being "Inbound" or "Outbound". This is something of a misnomer, as the lines for inbound calls and outbound calls are no different, and in most smaller systems are one in the same.

Banks of phone lines are referred to as "Trunk groups". Trunk groups are set up within the telephone system, and allow outbound calls to cascade from one line to the next, or fall to a specific outbound phone line (for, perhaps, auditing purposes). Trunk groups can be set up within a phone system to ring particular extensions, or to be attached to a button on a particular phone, for direct access to that line. All of this, in a digital system, is done electronically through software using programming. Meanwhile, the telephone company has instructions to group these very same numbers into a "hunt" group, which will be explained shortly. Hunt groups carry a nominal charge per line, but allow the lines to work together, so that (in this example) up to five people could dial (410) 111-1111 and not get a busy signal.

Let's use the example of a five-line system. Assume that the "advertised" number of your organization is (410) 111-1111. The telephone company has received an order from your organization to place each of your five lines in a hunt group, in the following order:

[1] (410) 111-1111
[2] (410) 222-2222
[3] (410) 333-3333
[4] (410) 444-4444
[5] (410) 555-5555

The lines in this example have been harnessed, such that the first caller will dial (410) 111-1111 and will actually connect via that circuit.

A second simultaneous caller to (410) 111-1111, rather than receiving a busy signal, will be [instantly and transparently] directed to your system through the subsequent circuit, which is (410) 222-2222 in this example. A third caller would be directed to 333-3333, and so on until the *sixth* simultaneous caller, who would indeed receive a busy signal.

Meanwhile, on your system, the lines are set up as such:

[1] (410) 555-5555
[2] (410) 444-4444
[3] (410) 333-3333
[2] (410) 222-2222
[1] (410) 111-1111

Note that this is the direct inverse of the first example. Just as the phone company has arranged that the lines will be available, 1 to 5, with 1 being the advertised number, your system is going to be configured to utilize every other line first, thereby minimizing the risk of the advertised number being utilized by an outbound call.

Some larger organizations, in fact, have two entirely separate trunk groups. One group is set up as an outbound group, another as an inbound group. The inbound lines are not allowed to be chosen by the phone system as available, thus minimizing the risk of an inbound caller receiving a busy signal due to all lines being used.

Some basic telephone systems, mostly those whom do not have programming, cannot utilize a programmed *Outbound* trunk group, per se. AT&T's Merlin system, for instance, and some of the smaller digital systems made under the Executone brand, have a series of plugs labeled A,B,C,D or 1,2,3,4 to indicate that the first line will be plugged into A, or 1. Subsequent lines go to higher letters or numbers. The lines are automatically selected based on which line is physically plugged into which port on the phone switch.

Most business telephone systems have some form of automatic line selection. Even very basic systems can detect that a line is in use, either by an inbound or outbound call, and automatically select the next available line. More advanced systems can actually be programmed to have separate trunk groups which are used by different extensions, or groups which can be accessed by pressing different programmable buttons on the phone station.

Inbound trunk groups, which will be discussed in greater detail later in the book, are a function of the local phone company, and will work with any phone system, regardless of configuration or brand.

CALL ANALYSIS AND AUDITING

One desirable feature of a system installed in a larger organization may be call analysis and/or auditing. If your organization has a number of users of the system, or relies on call center-like functionality, you can audit your call handlers using a statistical package.

Statistical packages give information as basic as call elapsed time, average call length, the phone numbers dialed by each station, and the time that calls were made. More advanced features can include such analytical tools as how many calls were abandoned prior to reaching a live person in the event that you use an automated menu system.

Usually, telecommunications systems which include this kind of functionality provide a software package for a PC which connects to the phone switch. The PC acts as a "logger" of the activity on the phone system, and records the information for display or analysis. Telephone systems which include this type of program are usually larger, or mid-sized systems, such as those manufactured by Telrad, Vodavi, or AT&T. Statistical analysis can also be of great aid in the event that you are experiencing seemingly "erratic" problems with your phone system or with a particular station. The logging software can sometimes point to reasons why a problem may occur, and is a good troubleshooting tool for phone system administrators.

COMPUTER INTEGRATION

Although not as much a "hot technology" as it once was, CTI (Computer Telephony Integration) can be infinitely useful for some organizations.

We discussed previously the options of programming of the telephone system through a computer. This is not necessarily the meaning of CTI. CTI implies that the phone system can be controlled by the user through use of a personal computer. The computer may include a program which keeps track of personal data, such as a phone list. By pushing an on-screen button, the computer can complete the phone call, instead of having to dial the number manually.

Additionally, CTI allows callers to take online surveys through the telephone system, and allows a computer to collect this information, and display and analyze the findings.

CTI is used extensively by telemarketing companies, using programs with large call lists, and pre-defined scripts which people read from screens. There are, however, more socially responsible uses for this technology.

VOICEMAIL AND AUTO-ATTENDANT

Even basic voicemail in the form of an answering machine will do wonders to streamline communications within your organization. These devices, around for decades, have become almost ubiquitous in the home. There are mixed advantages and disadvantages to using an answering machine in a business setting.

Answering machines have certain features which should be considered:

* Voice activated recording, so that you get the voice, not dial-tones, fax tones, or other misdialed calls
* Remote retrieval, so that one can receive the messages from home or a cell phone. It is best to also be able to change the outbound message from the remote location, in the event of a closure due to inclement weather, etc.
* Integrated Circuit (IC) operation. Use of tape answering systems is becoming less and less common, as time progresses. Tape answering machines use either one or two mini- or micro-cassette tapes. One records the incoming messages left by callers. Most machines which use tapes have a feature to use variable message length (this is tied into the circuitry for the voice-activated recording feature). This allows you to be able to store more shorter messages on the tape, than with a fixed 30- or 60-second message setting.

On older units, The outbound message may be on a tape, or on an IC memory bank. The advantage of the IC is that the solid-state memory will not tangle, or wear out like a cassette tape, which may answer the phone numerous times per day.

This is the primary disadvantage to the older cassette answering machines, in an office setting. The heads on the tape drive, and the tapes themselves, wear out more quickly, due to increased use, than those in the home.

There is a great flexibility with tape, though, over IC. Messages can be easily archived with cassette tape. IC units, without modification, cannot do playback into a cassette player. If it is important to record messages to archival tape, there may be some degree of difficulty doing so.

Similarly, units which take regular cassette tapes for outbound messages can have nicely recorded messages, with sound effects or music in the background, if desired. Most IC units can record only over the small, internal microphone in the unit. This may limit the options for your organization. Surprisingly, There are still many organizations, today, which have not made use of answering machine technology.

Perhaps of much greater utility than an answering machine is the voicemail option which may be available for your phone system. We are going to discuss a technology, primarily, referred to as Integrated Voicemail. Some telephone systems do not have the capacity for this type of functionality, others do. Voicemail is really a line which separates smaller-scale systems from larger-scale systems. As a system gets larger, integrated voicemail, which is usually manufactured for a specific model and brand of system, becomes an option.

Most telephone systems that offer an Integrated Voicemail package can operate perfectly functionally *without* the voicemail. The voicemail is an arm and hand of the system. The system can function without the arm and hand, but the appendage is of great help.

There are several different classifications of voicemail technology. There is the most basic of which, the ordinary answering machine. Answering machines are pedestrian technology, and require no lengthy explanation here. They operate in *parallel* to your telephone system, but are compatible with most all telephone systems which do not have voicemail of their own.

The telephone company also offers voicemail services, which are described in the next section.

A vastly popular application of a telephone technology is called IVM. Integrated Voicemail (IVM) can be, if managed properly, as indispensable as another member of your support staff. If set up properly, it will become an integral part of the day-to-day operation of your organization.

What does "Integrated" mean?

"Integrated" means that the voicemail component does not run in parallel to the telephone system. Consider this: Someone calls your office. The phone system rings four times, and nobody answers.

What happens? The answering machine, a separate device, has been listening to the ring on the line since the first ring. As programmed, it picks up the call, reads the announcement. The caller leaves a message, and hangs up. The answering machine resets itself, and the system waits for the next call.

Some higher-end answering machines are more complex, and offer such amenities as multiple mailboxes and basic menu features (press 1 for directions, etc.). These still, are operating in parallel to the phone system, and are not integrated.

Integrated voicemail means simply that the voicemail system is connected to, and can integrate with the phones. Voicemail, through programming, can transfer calls to extensions, or to the operator. People, without going to a "central location" such as an answering machine unit, can pick up their messages.

An integrated voicemail/auto-attendant system has the capacity to intercept up the call *prior* to any phone ringing. It is connected to the phone switch, and intercepts calls. A greeting is read, sometimes menu options are given.

VOICE STORAGE

The most obvious advantage of voice messaging is the ability to leave messages for individuals or groups of people. In this sense, voicemail is like a large stack of answering machines.

Mailboxes in IVM systems are usually set up in such a way that they attach to a telephone station. A light on the phone may blink or illuminate when a message is left for that extension. A button can be pushed, or an internal extension dialed, and a voice prompts the user on how to get their messages.

Settings, like the number of rings, the outgoing message, the password for the mailbox are set by following the system's female-voice prompts.

There can be, on most all phone systems, three types of relationships of the phone to the voicemail box:

1 to 1: There is one voicemail box to the phone
1 to 0: There is no voicemail box to the phone (the phone rings until the call is terminated).

There is usually no provision to have two voicemail boxes to 1 extension. Some larger systems can have 'phantom' extensions in the voicemail system "like 5555" which have no phone. By dialing a code, the users can check or leave messages for that mailbox.

At one organization that I worked for, there was actually a mailbox affectionately called "dial-a-date". Anyone could leave anonymous messages, identified only by their own voice that they wished to rendezvous with others in the office. It was remarkably successful!

Other, more legitimate uses of this technology would be a 5555 mailbox for a technology helpdesk. Calls could be left in the voicemail box, and picked up later by the technology "guru" for rectification.

So, what is important to know about IVM systems?

One aspect of the marketing of voicemail systems is a certain amount of "recording time". Advanced voicemail systems are actually computers with hard disk drives. The voice is digitized, and saved as files on the hard disk. The size of the disk determines the amount of time available for messaging.

Other, smaller systems, utilize solid-state memory, like computer RAM. These systems, which are much less expensive, do not offer the great number of voice-mail boxes and recording time. They are, however, much physically smaller, and are usually less expensive.

You will likely find that the type of voicemail available for your system is commensurate in features to the type of phone system that you have purchased.

Large phone systems usually have industrial-strength voicemail. Smaller systems usually have smaller-scale voicemail subsystems.

The other key factor of a voicemail system is how many ports are able to be simultaneously accessed into the voicemail system. This equates to how many calls can the IVM handle at one time.

Ports are like phone lines into the IVM. If the IVM has two ports, two users (inside or outside callers) can access the IVM at once, All others will get busy signals or will get dumped to a phone somewhere, depending on the settings in the phone system.

More ports equals a more expensive phone system, but can mean more user satisfaction, as well.

TIME AND DATE STAMP

Telephone systems usually have some capacity to know what the date and time are. A key problem with the older versions of the Merlin system were that it was not compliant with the year 2000. This is why there is a great abundance of this type of hardware available, relatively inexpensively, on the market today.

Time and date stamp mean merely that the time and the date are "audibly" stamped to the message. The automated voice reads you the date and time of the call when you listen to the message.

MAILBOX ACCESS FROM OUTSIDE THE SYSTEM

An invaluable feature of voicemail is the ability to check messages without walking to your phone, or having to go to work to get the messages. Being able to dial a code when the voicemail system auto attendant answers makes the system much more efficacious to those who use it.

Make certain that if your voicemail system is accessible from the outside world, that some level of password control is exercised.

VOICEMAIL FORWARDING

Voicemail forwarding means that the phone system will call a cellular phone, or home phone, or some other phone number outside of the system when a message is left at a particular mailbox.

It would seem to be annoying, but it really is of great help to your "roving staffers" who may spend very little time at the desk where their phones are located.

The phone system will receive and save the message from the outside caller, then will call the number pre-programmed into the system.

More advanced systems can be programmed to only forward calls on certain days or between certain hours.

AUTO-ATTENDANT

"Thank you for calling Conglomerated Furniture, inc, press one for sales, two for service, three for directions…"

So many organizations use these technologies today. Auto attendant, especially in the world of product support, is becoming increasingly pervasive, and advanced.

Where does this technology fit in your organization? Auto-attendant can be of great help to give pre-recorded messages about such things as mass or service times, if you're a church or congregation. If you have a facility, give hours of operation or directions. Accept donations? Instructions on how to give can be pre-recorded, too.

Rudimentary information such as the organization's website address, or physical address, can be added to highlight and increase visibility.

At our organization, for instance, all callers are given a reference to the website. "Thank you for calling St. Joseph Parish. Visit us on the web at www.stjoseph-parish.org"...A menu follows...

This kind of information can raise awareness, and can help to augment support staff by alleviating the need to give repetitive and oft-asked information to callers.

USING AUTO-ATTENDANT TO HELP LEVERAGE YOUR ORGANIZATION'S SUPPORT STAFF

One rule of thumb, however. Have no more than one level of menu to get to a live person. Call abandonment is a real problem, especially amongst people who may need remedial or assistance services. Not only is it frustrating to the uninitiated, it leaves potential and existing clients with the feeling, real or imagined, that your organization does not wish to communicate with them. Actual operators of some form should be able to answer calls in a professional manner, no matter what kind of work your organization does.

Telecommunications technology is a great thing, but misused it can alienate your client base and subterfuge your efforts to help those in need.

Most any organization could set up a scripted menu in the following manner.

> Welcome to A-B-C Organization. Visit us online at www.abc.org
> (Brief Pause)
>
> Press 1   at any time to reach an operator, or
>
> Press: 3—Directions and Address
>
>   4—Hours of Operation
>
>   5—Fax Number
>
>   6—Directory of Names

Incidentally, make "Zero" or drop due to use of a rotary phone go to an operator, rather than hanging up on the caller, as some organizations do.

As you can see, this organizational menu has no departments listed. There may well be one phone on one desk, and a lot of voicemail for each employee or volunteer. Yet, with this system, 25-30% of your routine call-types may be able to get basic information without speaking to a live secretary.

Did you notice that 2 did not appear as an option in the list? In at least one manufacturers voicemail system, all extensions begin with the number 2. This elimination of the "2" choice allows users to direct-dial extensions. If a caller knows that John is at 2205, she can directly dial 2205 at the beginning of the menu, and thus reach John without the intervention of a call-handler at John's organization.

Auto-attendant allows the switching of all calls, outside or internal, to any extension within the system, including the voicemail system for the purposes of leaving a message.

MULTIPLE DEPARTMENT SECRETARIES

Perhaps your organization has more than one department. Our organization has a parish office, a school, and a religious education office. Callers are presented with this choice of departmental options in the main menu when they dial the main phone number.

Each office has a larger phone which is manned by the secretary designated as the "call handler" for that department. Our school has only one secretary, who wears many hats, the other departments have a dedicated "greeter" individual who primarily greets walk-ins and answers telephone calls.

No matter how your organization operates, it is important to have a fast way to transfer calls from your call handler to others in the department, either to their direct extensions or directly to their voicemail boxes.

More often that you would think, a caller may be a salesman making a cold-call. In this case, the caller would be directed straight to the voicemail box, and the recipient could handle the message at his or her leisure.

DAY/NIGHT MODES

In some organizations, there are certain procedures which occur during the "day", that being normal operating hours, and "night", outside those normal hours. Installing an auto-attendant which can utilize a varied menu for "on-hours" and "off-hours", no matter when this line of demarcation may occur.

Usually, voicemail systems refer to this as Day-mode/Night-mode operation. Using and integrated voicemail system, this mode may be switched through the telephone system. Some manufacturers allow this mode to switch automatically, via a schedule utilizing the system's internal 24-hour clock.

A scheduled mode switch would be good for an organization whose personnel were consistently in the office during "normal" hours.

Some organizations have office hours which vary too much to effectively use a pre-programmed schedule.

Utilizing a manual-mode system with a programmed button on the Attendant console (The main "operator" phone) allows your office staff to switch from the "in the office" settings to the "out of the office" settings.

LANGUAGE AND PROFESSIONALISM

When utilizing the features and benefits of Auto-attendant, it is good to exercise two practices. If possible, someone who is comfortable with their voice and the idea of recording should be responsible for recording the guide messages in the auto-attendant system. Auto-attendant systems need not change that dramatically over time, but try to get someone who has a pleasant voice, and who, more importantly, is going to remain with the organization for a presumed long period of time to do the messages.

A patchwork quilt of voices is a very puzzling experience for most people who are uninitiated with your system, and re-recording all of the messages within the system every time you need to make a change is time consuming and impractical.

Likewise, it is a good idea to script the wording of your auto-attendant menus prior to beginning the recording process.

How does the recording of auto-attendant messaging work? There are usually announcement boxes to which recorded messages are saved. One such recorded message would be the top-level message. Other recorded messages may be the one played containing directions, or time of operation.

A script would be a multi-page document, including all wording of all recorded messages. 'Soft' items such as pauses, or intonation, could also be specified, so as to maintain aural continuity when changing the wording of messages. How detailed you get depends on your organizational objectives and preferences.

Many organizations standardize their staff voicemail messages, so as to avoid confusion, and to eliminate the nervousness or anxiety about leaving the message. Our organization did this, and found it to be highly successful.

An instructional sheet was sent out to all of the voicemail users. The instructions clearly and succinctly indicate how to enter the voicemail system, how to get through the menu to the place where the messages can be recorded.

Beyond that, there are scripts for two message types. One is for the regular outgoing message, which includes instructions for getting to an operator, and leaving a message with one's name and number.

The second is a message for the "I'm on the phone, and the line is busy" situation. This message, worded the same way, indicates that the destination is on the phone, and to leave a message with the name and number.

With regard to your auto-attendant, specifically, your messages should be worded carefully, and should be simple enough to understand for someone who is not versed with your system. Avoid superfluous mailboxes, as an unending list of choices both intimidates and confuses users of your system.

MAIL BOXES

Mailbox 9000	"Welcome to A-B-C, a non–profit organization. (1 second pause) Visit us on the web at www.abc.org" (three second pause)
Mailbox 9001	"Press one, or remain on the line for an operator. (1 second Pause). Press three for directions, four for hours of operation, or five for instructions on leaving a clothes donation."
Mailbox 9002	"Directions to our facility. Go to main street, and take a right…"
Mailbox 9003	"We are open from 9 to 5 Monday through Friday. and 10 to 5 on Saturdays"
Mailbox 9004	"Please leave any unopened food or personal supplies, or clothing at our offices on main street any day during normal business hours".
Mailbox 9005	"Our offices are now closed. Offices open at 9 AM, Monday through Friday, and 10 AM, Saturday. (2 second pause)
Mailbox 9010	"If you require assistance, please dial zero for the operator. Dial pound to hear the menu again"
	Press 3 for directions, four for hours of operation. If this is an emergency, please remain on the line, and you will be transferred to our after-hours answering service"

Programming would initiate a set of events to occur along the lines of the following script.

During the day, when the phones were in "Day mode", auto-attendant would initiate 9000, then 9001, when a line was picked up. Even though two message segments would be read, the user would not be aware of the break. The system would wait for user input. If no such input occurred after 5 seconds, 9010 would be played. If still no input occurred after 5 seconds, the call would be transferred automatically to extension "0", the operator telephone.

If the user pressed 1, they would be directed to extension "0". If they pressed 3, they would hear mailbox 9002, and be returned to mailbox 9001. Likewise, if they pressed 4, they would hear mailbox 9003, then be redirected back to mailbox 9001. Finally, if they pressed 5, they would be directed to mailbox 9004, then be redirected back to mailbox 9001.

During the off-hours, when the system was switched by the attendant to night-mode, the system would do a similar set of events. Auto-attendant would initiate 9000, then 9005, when a line was picked up. As with day-mode, the system would wait for user input. If no such input occurred after 5 seconds, 9010 would be played. However, if nothing happened, the user would be directed directly into the voicemail box for extension 1010, the attendant console, which is the extension which rings when the number "0" is pressed.

If the user pressed 1, they would be directed to extension 1010's voicemail. If they pressed 3, they would hear mailbox 9002, and be returned to mailbox 9005. Likewise, if they pressed 4, they would hear mailbox 9003, then be redirected back to mailbox 9005. Finally, if they pressed 5, they would be directed to mailbox 9004, then be redirected back to mailbox 9005.

These scripts should include whether the mode is operational on the day mode or night mode, and the "mailbox" (announcement storage) number, so that they can easily be changed.

Your organization may also wish to specify, using a log, when the messages were last updated. A log could easily be attached to the front of the binder where this information is stored.

VOLUME

The volume of the voice is important. Some auto-attendant systems are very sensitive, and as such can seem very loud and tinny to the caller. By contrast, others are too dull, and seem too soft to understand the messages. The recorded messages usually come through OK, but make sure that your recorded messages are being monitored from outside of your system, for volume level.

Volume can hinder communication, too. Speaking too loudly can destroy rapport between yourself and the caller. Even if the caller is hearing your voicemail message, this can work for, or against you as an individual and your organization.

DICTION

Just as appropriate volume is crucial, speaking clearly and succinctly (perhaps artificially so) is important. Users of your system will hear your voice, and must be able to understand what you are saying. Speaking directly into the handset for the recording of messages will help. Additionally, word choice is very important. Without sounding "stiff", your word choice on your message will affect your callers' perception of your professional level. Using slang or colloquialisms in your messages, accenting words, or sounding gruff can work against you and your organization.

LANGUAGES, THE DEAF, AND ROTARY PHONE USERS

What do these three people have in common? José can't speak English, Marguerite can't hear at all, and Aunt Polly has a rotary telephone, (property of Bell System, not for sale.).

If your system is not friendly, none of these individuals can get through to a live person.

Sometimes the voice of your organization isn't audible, at all. Making your system friendly to people of other nationalities, hearing impaired individuals, rotary phone users as well as other handicapped individuals is of great importance when running a benevolent organization. Perhaps your largest contingent doesn't speak English well enough to traverse your phone system. This undermines your ability to reach and communicate with your client base.

Make your system friendly without hindering usability. If you have a significant population of Spanish individuals, start your message with an option for an alternative language option. "Press 2 for Spanish", for instance (said in Spanish, of course).

The hearing impaired should always be considered when planning a system. There should be a separate TTY number if there is no provision for alternate communication through your telephone system or auto-attendant.

Pressing "Zero" should <u>always</u> bring the user to an operator, or barring an operator, some live, knowledgeable and friendly individual within your organization. Consult the person who will become the default "operator", and make sure that this regular distraction is acceptable to them.

Rotary phone users, too, should be considered when programming and setting up your auto-attendant system. Put your phone system to a simple test:

You need not have a rotary phone to do this. Call your main phone number, what happens if instead of pressing any number, you simply wait for something to happen.

What happened? Did the menu repeat itself indefinitely, or did you get the hang-up? If your system is set up in a friendly way, there should be a brief pause after the end of the menu, and a choice to repeat the menu.

If nothing is activated by the caller, the auto-attendant should transfer the caller automatically to a *live person for assistance*. If your live person is the operator, or secretary, it should always ring that phone. Putting someone in voicemail, or hanging up on them is a good way to kill goodwill within your organization.

What if one of your staff is hearing impaired? To aid those who are hard of hearing, the ADA specifies requirements for a business telephone systems' ability to work with those who are, to some extent, hearing impaired.

The way in which most of these systems aid those who are hearing impaired is to provide an optional attachment to the telephone station. These devices connect to a connector on the phone, or in-line with the receiver. One company makes a battery-operated amplifier which can significantly increase the volume to the handset. Even if your staff is not hearing impaired, devices which amplify the sound and headsets which allow less ambient noise to interfere with the phone conversation can be of help in a noisy environment.

AUTO-ATTENDANT DIRECTORIES, EXTENSION PRIVACY, ETC.

Depending on the nature of your organization, it may be desirable to solidify a policy on giving out direct extensions to outside callers. The outside extension is a key to the door, in a way. Once this number is given out, it becomes a way to get directly to a user.

Phone system manufacturers realized this, and have created a number of ways to handle this situation. Many modern phone systems have options to forward *external* calls differently than internal calls. Some users would prefer to have all external calls go directly to their voicemail box. This is different than Do Not Disturb, in that it is programmed directly into the phone system, and the outside, direct calls (as opposed to outside calls which are transferred by the operator) never ring the phone station.

Another way to handle outside callers is to give them access to an automated directory. Directory services are a specific subset feature of an auto-attendant system. There is some option for a directory function with most newer auto-attendant systems.

Telephone directories are like the phone book, for the users within your phone system. Users can be given a choice in the main menu (press five for the directory). Pressing five initiates a pre-programmed script which asks the user to press the first couple of digits of the first or last name. The phone switch then matches these digits up with names programmed into the system by the system manager. Without giving the extension information, the directory can either transfer the call to that extension, or transfer the caller to the voicemail box of the recipient.

Some organizations advertise the presence of these menus, other organizations do not. Just as with extension direct dialing, it may not be desirable to advertise this feature's availability. As with any feature, the directory can be used with or without the presence of a spoken guide indicating its availability.

Forwarding calls

One programming aspect of more advanced telephone systems is their flexibility to handle different kinds of call-types.

There are internal calls, calls from one station to another within the system. Some systems have a distinctive ring for these kinds of calls.

There are outside direct-extension calls, where the person dials the extension. There are also transferred outside calls, where the outside caller has spoken to an operator, and has been transferred to your phone.

Telephone systems can be programmed to handle these calls differently. Presumably, a secretary would screen callers, prior to transferring the call to your phone. Some calls he or she may transfer directly to the voicemail box, others may be transferred to your extension.

Direct calls to your extension are not filtered, so phone system manufacturers sometimes put a facility into the system to allow these calls to be handled differently, depending on the preference of the user.

One key difference lies not in auto-attendant, but in the voicemail system. One manufacturer allows no less than five different greetings to be programmed for each mailbox.

There is a greeting for inside callers. This greeting may be used to tell inside people that you are out of the office today, or are going on vacation next week.

There is a greeting for outside callers, which may be worded differently, or may not give out personal alternate-call numbers, such as cell phones.

There is usually a message for busy internal calls, and a message for busy for outside callers. This is the message which is played if someone calls you and you are already on the phone.

At least one phone system manufacturer allows for a "special" message to override these other messages at will (like a message indicating to all users that one is out of the office), without having to erase your carefully worded greetings. The telephone system will vary depending on the brand of voicemail, but these powerful features allow for a cohesive system which allows communication to flow in appropriate directions.

Section 3:
Installation,
And Everything After

KEY PLAYERS

The first key player is you, the owner/operator of the system. You are the first party, because you are designing, and ultimately responsible for the satisfaction and efficacy of your communications system.

* The Owner
* The Interconnect Contractor
* The Phone Company
* The long-distance provider
* The manufacturer

Your Primary Role:
> Owner
> Final decision maker
> Requirements Designer
> Signer of the checks, payer of the bills

You interact with/pay:
> Telephone Company for local and long-distance services
> "   "   for business services having
> to do with your phone lines
> Long Distance company for Long-distance service
> Interconnect Contractor for Phone Systems and Telephones
> "   "   for Installation of wire, phone system

The second key player is the local telephone company. The local telephone company is going to give you a certain amount of "head-end" hardware, which is going to allow you to interconnect your companies equipment to the network. The phone company will also give you services, at an additional price. Services like "Caller-ID", which allows you to see the name and/or phone number of the call which is coming in. There are other services available, like the ones described in the "Hunt group" section, and voicemail.

Prior to agreeing to setting up telephone services with the phone company, ensure that you do not already have the functionality built-into your telephone system. Ensure also, that the functionality from the phone company is compatible with your phone system. For instance, even though they may have alphanumeric displays, some older phone systems will not be able to work with Caller-ID services.

Telco's Role:
    Provides Local dial-tone
    Provides all of your telephone numbers
    Provides your telephone listing in white/yellow pages
    Provides configuration of your phone services

Interacts with/Bills:

    You, for Local phone service
    "   for Extended services for your phone lines
    "   for Directory listings
    Interconnect Contractor, defines standards for workability
    of third-party equipment with the phone network

The third key player is the long-distance provider. The long-distance provider is the least important of the key players, because their services do not have any integration at your level with other players in the arrangement.

Another key player is the Interconnect contractor, or contractors. These individuals are the group whom is contacted to design and install your telecommunications system. This group is not required, should you have a volunteer or staff individual who is versed in installing the equipment for your organization.

Interconnect Contractor's Role:
    Provides equipment for installation
    Provides installers and service personnel
    Provides expertise and advice for design
    Provides expertise for local/national codes and laws

Interacts with/Bills:
> You, bills for design/installation service
> " , contracts with for system installation
> Phone company, works with to get phone lines installed/activated

The final key player is, of course, the manufacturer of the phone system. This company made the equipment. They interact with the phone companies of North America, by adhering to the standards set forth by the groups who set the standards for the North American telephone system.

Role:
> Provides equipment to contractor
> Provides knowledge, training, support to contractor
> Provides knowledge, manuals to owner
> Provides support, repair, and parts for service contractor

Interacts with:
> Contractor, bills for equipment
> " provides support, training, parts, knowledge

INTERCONNECT EQUIPMENT VENDORS/ CONTRACTORS

An interconnect contractor is a company who installs phone systems. They hire individuals with experience in the trades having to do with pulling wire in walls, and dealing with codes regarding electrical and data-communications wire.

Most urban areas will have the service of at least a few interconnect contractors. These companies usually work with businesses to install phone systems, data-networking wiring, and PA systems.

Prudence advises to utilize a "real" telecommunications interconnect contractor, rather than a company whose primary interest and expertise is in data networking. There are certain similarities and overlaps between phone systems and computer networks, but the expertise is varied.

It is also a good idea to make certain that your phone system has a certified installer program. This can be found out easily through the manufacturers support line, or website. Phone system manufacturers keep lists of authorized dealers and installers, and can direct you to competent installers and service providers in your area.

The relationship with the interconnect contractor need not end with the completion of the installation. Most interconnect contractors are also certified maintenance providers for the systems that they install. These organizations can provide in- and out-of-warrantee service for your phone system.

Choosing a contractor can be difficult. Someone once said that any contractor is part craftsman, part businessman, and part magician. This is true. The contractor has a vested interest in doing a good job, although not doing such a good job that (1) the job loses money, and (2) the contractor has no opportunity for changes to your orders, and service down the road. Most contractors' bread-and-butter is post-sales support of phone equipment, and of course, service contracts.

As with any contractor, you should look for clues as to the level of experience which the contractor may or may not have.

Is the contractor insured, do they have equipment and trucks. Not to generalize, because there are some wonderful independents out there, but personal vehicles with ladders strapped to the top, and rented tools from the hardware store are usually a sign that a great deal of supervision may be necessary.

A good contract will provide, in due course, the following things:

* Sketches of the work to be done
* Plans detailing the time frame, and work schedule
* A Materials list, itemizing expected expenses as line-items showing brands, quantities, and cost.
BE WARY of any "or equal" clauses, which permit the contractor to use their excess parts from the last job, or the cheapest alternative that they can find to do the job.
* The payment terms, sometimes done in 1/3's. One third up front, one third after a certain mile-marker, one third at the end of the job, which is signified by the owner's approval and sign-off.
* A proof of insurance for workers' compensation, liability and damage, and automotive.
* A rider, stating clearly, that all changes, whether or not they affect the cost of the contract, must be submitted and approved in writing.
* a rider, stating clearly, that all walls, ceilings, floors, and other architectural items, will be returned to the original condition as prior to the installation.
This, of course, should represent the least of the works of a contract. Your contract may have some other items, depending on your organization's specific needs.

THE PHONE COMPANY

If the interconnect contractor is outnumbered, it is probably in the area of the third-party telephone company. Since the re-organization and federal regulatory changes in telephone services providers, there are a great many people selling phone service.

My advice? I used an independent phone service provider, a national company, for three years. My personal experience with my account of 26 phone lines was this:

* The individual service experience was superior to Verizon, in that there was a particular individual whom I called when I had trouble.
* The overall experience was inferior to Verizon. Problems that manifested themselves and were beyond "no dial-tone" were never dealt with in a satisfactory way or completed.

For instance, we had a fire alarm with an automatic dialer. The fire alarm, as code requires, had two dedicated phone lines connected to it. Should one not be able to complete the regular self-test (an automatically generated call to the monitoring station), The second would dial the monitoring station, and the system manager (me) would get called to indicate that a problem occurred.

UL, the certifier of Fire-Alarm monitoring stations, requires that the system manager be contacted WHEN THE SYSTEM FAILS. I would get called at midnight, at three in the morning. Whenever the self-test would fail, I would get the call.

This was an obnoxious situation. Because there was technically nothing wrong with the dialer module itself (the electronic system which makes the call), the alarm system maintenance agreement would not cover the repeated service calls to the system.

The problem lay in the way that the dialer worked with the phone company's equipment. Because our local dial-tone provider used its own equipment, not Verizon's equipment, there was some degree of difference in the way that the actual tones made by the touch-tone dialing system were handled on the phone company's end. The speed seemed to be the problem. Our dialer, even though it was in the satisfactory range, was not in the range for the third-party phone service provider's equipment.

Their engineers, over a two-year period, were unable to resolve this problem.

My advice: The savings for phone service is vary little from one phone company to another is very small. The added soft-cost for a small-business, or not-for-profit users is very high. I would stay with the primary phone company, if I were asked to make a decision.

TELEPHONE COMPANY SERVICES

The telephone company, as we discussed earlier, will service certain aspects of your phone system:

* Billing problems with your phone lines
* Operational problems with services provided by the phone company, such as caller-id
* Discontinuity with regard to dial-tone service
* Problems with directory listings, 4-1-1 listings, etc.
* All outside wire, either in the ground or on the pole
* Problems which occur at the central office (the phone company facility)
* In most cases, wire installed on the outside of your building
* Wire installed inside your building, up to the point of the "Entry facility". The Entry Facility is the hand-off to the owner's equipment.
* The Demarcation cabinet equipment, and the primary grounding equipment(The grounding equipment which protects the phone company's lines from damage due to malfunction in your equipment, or lightning).
* Wire inside of your building, up to the "jack in the wall" if you have an extended agreement with the phone company to perform this kind of extended service.

   The Interconnect service contractor, if you have one, will usually be contracted to service:

* The "House Wiring". That is, all of the wiring which is installed beyond the Entry Facility.
* The Patch panels, and blocks which are used with your system
* The telephone switch, telephone stations, and associated hardware
* Moves, Adds and Changes to your phone system

Basically, the Interconnect service contract, if it is comprehensive, will assign all responsibility over the physical phone system operation to the contractor which is not handled directly by the phone company.

THE NATURE OF THE INTERCONNECT EQUIPMENT VENDOR RELATIONSHIP

Should you install a new phone system, expect to do at least four years' business with your interconnect contractor. Unless "irreconcilable differences" occur, you will likely be involved in a service contract of 1 to 2 years, the warranty of 1 year, and the initial design/installation of 12 to 18 months.

Many businesses expect to get a decade of service from a phone system. This is not an unreasonable expectation. A telecommunications system should give at least 10 to 15 years service. During this period of time, replacement parts, additions, configuration changes, and regular service will almost certainly be necessary.

It is handy if the interconnect contractor is retained, because that firm has likely installed many similar systems, and is familiar with your installation.

Unlike computer system vendors, and computer-networking installers, telecommunications systems contractors tend to be long-term relationships. If for no other reason, this is due to the proprietary nature of the physical aspects of your phone system. *Factory new* parts and additional equipment are usually available only through the authorized dealer, who is likely the company who installed your system.

SYSTEM PRICING: HOW SYSTEMS ARE PRICED, FROM THE ONSET OF PURCHASE, SERVICE CONTRACTS, MORE EXTENSIONS, MORE CABINETS, AND MORE FEATURES

The initial cost of the phone system is based on a number of factors:

Wiring and Infrastructure: There is a per- frame-to-station cost of wiring your building. The interconnect contractor will call each piece of wire which runs from the telecommunications equipment room to an office as a "drop". Each telephone station requires a "drop", and if you use the schema described in appendix I, there are two drops for computer networking purposes. This would mean that each office would be multiplied by 3 for the purposes of calculating wiring costs.

Additionally, infrastructure equipment, such as patch panels, punch-down wiring blocks, grounding equipment, racks, labeling, etc. factor into the infrastructure cost.

Telecommunications Equipment: The telephone switch is one of the highest cost item. In a smaller system, there will be one cabinet, one power supply, and depending on your system, a series of add-in modules or cards.

Phone switches vary in price, significantly. Most are sold depending on the capacity of phone company lines which can be connected, and the capacity of telephones which can be connected.

One primary function of cost is the ability to connect voicemail to your system. Voicemail ability significantly affects cost.

Outside of the actual cost of the phone switch, there will likely be certain add-in modules or cards necessary to make a functional system. Add-in cards will likely include a required processor board, a card for the voicemail/auto-attendant interface, cards for phone lines to the outside world, and cards for internal telephones.

Telecommunications systems also include the actual telephone stations. Most manufacturers make a variety of types of telephones. Some less-expensive models will resemble ordinary telephones, and will have no, or very few programmable buttons. More expensive models may have visual displays of some form, more programmable buttons, speakerphone, or other extended abilities.

To estimate your costs, count the number of desired stations, and add 1/3 to this number for the purposes of the phone switch capacity and future planning.

Voicemail/Auto-Attendant: These add-on systems will usually purchased as a unit from the manufacturer of the phone switch, and may include an additional cost for an add-in card or other attachment module.

Sometimes, there is an incidental cost of additional phone lines to accommodate the functionality of the voicemail or auto-attendant. This topic should be discussed with the Interconnect Contractor prior to purchasing a system.

POST-INSTALLATION COSTS

Additional telephone stations are a step-stair expense. One additional phone may be limited to the cost of the phone itself. As of this writing, the cost of a PBX phone ranges from $100 to $500, and up, depending on manufacturer and the type of phone required.

At some point, the phone system may require another add-in card to handle more phones, which may require the purchase of additional hardware. Add-in cards are more expensive than phone stations, and their price varies significantly depending on the manufacturer and the system.

At some point, the phone switch will be unable to handle more cards or modules. Some system can be linked together, but this requires either an expansion cabinet, or another primary cabinet and power supply. There is also an installation cost involved with this option.

Not getting to this third point should be the joint goal of you, as the owner, and your interconnect contractor. Option three becomes your hypothetical "$4500 telephone". The phone, at $500, the card at $800-1000, and the additional switch, at $3000+, will likely negate the need for the additional station. This situation should be avoided if possible. The best way to avoid a situation where an additional switch or expansion cabinet need be purchased is to figure out the outside number of phones which need to be installed, and to look at the cost involved with installing this number of phones. Have a frank discussion with the contractor about how this would work, and what the cost would be at today's prices to add the next phone beyond capacity.

The Service Contract

Whether or not you decide to go with a service contract depends on your level of dependence on a working system.

Like many other industries, the nature of phone system service companies is that the service contract customers get the priority over the non-service-contracted customers.

Perhaps your organization cannot afford the service agreement. Consider a limited agreement, or a time and materials arrangement with the interconnect provider. Also, because you're a non-profit, use this technique to get a better price. Who knows, sometimes it really does work. Schools and churches, especially, may be able to utilize their communities to find parents or congregants in these industries who are interested in ways to help their communities.

What does a service contract look like? Service contracts range from one page arrangements to multi-page contracts.

Key areas of a service support plan include:

Normal hourly rates as of the time of the service contract

* Time rates for normal hours service
* Time rates for after-hours service on a work day

* Time rates for after-hours service on weekends if different, time rates for holidays

The Plan which is chosen should be specified as to when your service contract is in effect.

Most service vendors will offer various options for service levels. Higher priced service agreements offer service 24 hours-a-day, seven days a week.

Other, less comprehensive agreements are valid five days a week, and allow a certain amount of delay for holidays and weekends. Your agreement should describe the different offered options, and specifically indicate which option your organization chose to go with. Another good question to keep in mind is regards the delay time between reporting a call and getting service personnel on site. Is this time one hour, four hours, all weekend, and how important is this delay, in as much as the interruption to your business operations.

An attached list of what equipment or wiring is covered

This list should include a detailed list of the parts of your system, and the number of phones of each type specified.

An explanation of who pays for replacement parts on covered equipment should also be specified.

A Service Level Agreement, SLA

How long will it take to get service if an outage occurs
during work week, on the weekend, on a holiday?

Definition of an outage

Worded definition of what defines a system outage.

The term of the contract

How long is this contract to last: One year, three years?

Is there a penalty to your organization or to the contractor if the contract is broken? Is there a provision to break the contract without penalty if the terms are not met by the contractor?

Cost terms

The cost of the contract over its term, and how the payments are to be made (e.g. annually, semi-annually, quarterly). Perhaps your agreement specifies a discount rate because of the payment terms that you specified.

Regular Maintenance

If your system is to receive regular maintenance, the terms of this should be specified in this agreement. The items which receive maintenance and the nature of the maintenance should also be attached to the contract.

Is there a way to back out of the agreement if regular service is not being provided?

Be wary of support agreements which are worded in such a way as to negate service, or are clearly in the best interest of the vendor, at the expense of your organization. Service agreements are legal contracts, but they are also insurance policies, and should be treated as such.

Look at the cost of the service agreement versus the cost of hourly service. Ask the vendor also if there is a policy on the priority of time/materials (non-contract) customers versus service-agreement customers.

In-Warrantee Service versus Out-of-Warrantee service

It can be a good idea to set up a service contract at the point in time when an installation agreement is being negotiated. The service contract is the 'bread and butter' of the interconnect industry, and the contractor may better the deal to get your signature on a multi-year agreement.

Prudence dictates that you should never sign a 24 or 36 month agreement in general with an unknown vendor. However, sometimes it can be to your advantage if the agreement guarantees a level of service over the term and a fixed price on parts and, by its nature, labor.

Make certain, though, that the term contains wording acknowledging the manufacturer and contractor's warrantee. The service contract term should begin *after* the term of the warrantee is expired.

If money is tight, some contractors, to alleviate the cost, will set up a multi-year agreement and spread the cost of a three year service agreement over the three-years plus the one or two years that the equipment is under warrantee. This gives the net effect of a five-year service contract with a spread-out payment schedule, thus reducing the cost of the service contract.

FUTURE SUPPORT, AND THE LIFE SPAN OF THE SYSTEM

Make certain that part of the service contract involves a guarantee for the future cost of parts for your telecommunications system. This agreement should be in writing, and should cover all replacement parts and additional parts for a term of time, three years, for instance.

When dealing with a new phone system, it is of import to ensure a time frame for the response delay between calling in a problem and getting a service response, and a time frame for getting parts from the manufacturer.

Life-span is determined, in large part, by the amount of time that replacement parts will be available through the manufacturer. When the parts supply dries up, you are relegated to refurbished, and other secondary markets for replacement parts. Ensuring, in writing from the manufacturer (or, if this not possible, the Interconnect Contractor), a minimum number of years that parts will be available is both prudent and advisable.

ON PROPRIETARY DESIGN

For reasons stated earlier, it is important to maintain a relationship with both the vendor and manufacturer of your phone system. Phone systems, are by nature, proprietary. Proprietary in the same way that a car is proprietary in design. A GM drivers' seat will not likely fit into a BMW car. In the same way, Brand A telephones will not work with Brand X telephone systems. Handsets, labeling and designation strips, all parts for that matter, are tied to the manufacturer, or to a third-party vendor of replacement parts which will work with your system.

Unlike the personal computer, whose interface was standardized by IBM's PC and the Microsoft DOS and Windows interface, telephone systems all operate in a different way, using different types of systems. Every aspect that has become standardized in the world of personal computers has remained de-standardized in the world of telecommunications. This includes user interface, function code operation, data bus, interface, and architecture.

In the same way, knowledge and experience of one manufacture's phone system is not necessarily useful or accurate for any other manufacturer. Operation, feature sets, and architecture vary significantly from one brand to another.

MANUALS

Because of the proprietary nature of telecommunications systems, operational manuals are of key importance to those servicing and managing telephone systems.

Manuals for phone systems are like the manuals for automobiles. Most every car includes a basic user manual, which describes how to use basic functions, and 'troubleshooting'.

Additionally, there is a real manual, for servicing the car. The manual discusses how to take the car apart without breaking key parts. How to service the car, and how to maintain the equipment.

Phone system manuals are the same way. There are usually "cheat-sheets" for the users of the system, and manuals for the service provider. These manuals should be (1) purchased by the owner, and (2) kept on site, wherever the phone system is kept.

This ensures that the phone system is able to be serviced, and that should knowledge be needed, that it is readily available.

When dealing with used or refurbished equipment especially, there is a tendency to not receive all of the manuals. However, there is a tendency for vendors and contractors to fail to turn over manuals to the owner unless the owner specifically asks for these manuals.

Operational and attendant manuals (the thick ones which discuss programming and technical details) are in high demand, and are not usually part of the refurbished equipment packages. It is a good idea to understand the availability and cost of the operations manuals for any equipment you wish to purchase.

REFURBISHED HARDWARE

We've dealt with the tough chapter of the areas of expense with regard to a telecommunication system. Let's talk about how your non-profit can reduce, or eliminate some of these operating costs.

Factory New and replacement hardware usually must be purchased through a defined channel. As and end user (owner) of a phone system, you usually cannot call a phone system manufacturer, and purchase one piece, like a handset. Usually, the manufacturer sells parts through a network of dealers, who service and support their equipment.

In reality, it is difficult to outright purchase parts for most systems through the dealer network. These parts are usually only available to the dealers via a trade-in policy. If you have a phone, or a card which goes bad, you need to trade in the existing part to get the replacement.

Refurbished hardware is the other option. Buying refurbished cards and telephones through a third-party is a viable option if you do not have a service agreement with a vendor. Looking on the internet can be a good way to find these clearinghouses.

Like purchasing other replacement parts, phone system parts are usually identified with product and model numbers. Knowing exactly what you need, and the part number is key to being able to purchase through a refurbished-product channel, as the salespeople may not be particularly versed with your phone system, or its features.

If you choose to do some of your system's maintenance yourself, make certain that part of your installation agreement allows for access to the software or overlays which are usually required to program the phone system. Without these programs, and some knowledge of how your phone system works, you may find that even the most rudimentary repair is fraught with difficulty.

DONATED AND USED EQUIPMENT

The other realm of used equipment is the donation. Many companies donate computers and phone systems to not-for-profit, because they upgrade, and like the tax advantages of the donation.

If managed well, this can be a great way to get a good phone system. Companies who are going out of business, moving to larger quarters, and upgrading their phones are often looking for a simple way to remove the old phones and equipment.

The same caveat goes for donated and used hardware. As with any equipment donation, speaking with the owner or system manager (or both) is advisable. The manuals, overlays, and supplementary materials should be obtained, if possible, from the original owner.

SYSTEMS YOU CAN INSTALL YOURSELF

Certain systems are easier to install and remove than others. Generally, smaller phone systems of less than 25 or 30 terminals are easier to remove than newer systems.

Installation of a telephone system is not the easiest thing in the world to do. Systems, because of different generations of equipment and different ways to adhere to standards, are all wired differently.

A piece of advise is to have a networking specialist, or someone versed in phone system installation remove, and install your phone system. Some of the hardware belongs to the telephone company. Removing or damaging this hardware will cause problems for the owner or tenant of the building where the system was originally installed.

Likewise, house cabling for the system belongs to the building. Removing the wiring facilities from the building will cause untold problems for the next tenant.

If you are not familiar with network or telecommunications wiring, it is your best bet to locate a professional, or solicit the aid of a knowledgeable volunteer or acquaintance.

Complex systems, especially the old Bell System business phones should not be removed by the uninitiated. The wiring is complex, and is often un-documented at the owner's facility. The easiest systems, which are generally available, to remove and reinstall are the AT&T Merlin, Spirit, and Partner systems.

In personal experiences, I have found that smaller Vodavi- and Executone-manufactured systems are easy to remove as well. A good tip for anyone is to take a photograph of the phone system closet.

REPLACEMENT DESIGNATION STRIPS

Whether you, or a volunteer, installs your phone system, there is one aspect of the system which will almost certainly need to be replaced. The designation strips for the telephones are customized to the system. Many times, these strips contain people's names, or names of departments—names of the old ownership and the old organization.

Usually, companies who sell refurbished parts can handle getting "consumables" for your phone system. These include replacement (original) designation strips, handsets, and cords for the system. Usually, these are the pieces that need to be replaced.

There are also a few companies (look up the word "Desi" on the internet) who sell printer-friendly blanks for phone systems. You indicate what brand of system your organization has, and they send you a package of 25 or 50 blanks in a package. A program, which is proprietary to the company's product, prints the writing in the right place on the page to fit the template.

These are very professional looking, and relatively inexpensive and easy to implement for a small organization.

SERVICE AND TROUBLESHOOTING

Performing your own service and troubleshooting can help to offset the cost of running your telephone system. Simple programmatic and technical difficulties can usually be corrected (or at least diagnosed) by the owners and users of a simple telephone system. This can help to offset a lengthy diagnosis charge from the Interconnect contractor. Sometimes, it can be useful to try to solve the problem yourself before calling the service department. Like with any appliance, there are simple things which can be checked by the owner.

The service provider would ask you certain questions, for instance:

* Is the entire system "dead"
* If the system is OK, can anyone dial outside of your organization
* If the phone system is OK, and the phone line is OK, Is the particular phone in question seem to be "dead", or can it dial internal extensions
* If the phone is not dead, check the programming configuration. Usually this is a programmatic problem.

This framework, and all troubleshooting frameworks, work from the big picture to the details of the particular system, not the other way around. Picking up a phone and not being able to make a phone call does not indicate that the "entire phone system is down!". Perhaps the phone is not allowed to make outbound calls.

For simplification, and ease of repair, some particular individual who is (1) familiar with the programming of the phone system and (2) regularly available should be in charge of troubleshooting and correcting the problems with the system.

That same individual should be in charge of maintaining and making changes to the telephone system. Should that individual not be regularly available, a secretary, or other support person who is regularly on-site should be in charge of maintaining the system on a day-to-day basis. This person should be familiar with how to move extensions, make changes to the auto-attendant, and voicemail system, and basically be familiar with "simple" problem solving.

Incidentally, to be able to effectively troubleshoot problems with your telephone system, you will need some basic tools. Take a look at the section about a "basic telecom toolkit", which will help to guide you in a basic direction.

INSTALLING WIRE

Many network contractors charge between $125 and $200 to install the wire necessary for a new phone jack for your system. This cost can go up if your building is complicated, from a wire-installation standpoint. Old houses, older buildings in general, buildings without drop ceilings or drywall, and buildings with stone or block walls are likely candidates for a "difficult" wire installation.

Wire installation is usually done by a "structured cabling" installer, or contractor. These companies have the equipment to correctly, and safely, pull and install communications wire. They are familiar with techniques regarding multi-floor offices, and testing of the wire which they install. Can you do it yourself? This book will not guide you in that direction, but there are many educational opportunities for people who wish to install network cabling. Generally, while pulling wire in a newer building can be reasonably straightforward, this activity should not to be taken lightly. There are a great many codes which define working around electrical cable, fire stopping, and correct installation technique such as labeling, and materials, one should be familiar with wire installation before attempting to venture into this realm.

PHONE GUIDES AND DIRECTORIES

There is a great little tool for any organization—a phone directory. I use in my organization a half-piece of paper, printed on both sides. The paper is laminated where the location calls for it, and is small enough when cut down the middle to fit under the telephone stations that we use.

The front of the paper gives the phone numbers, the back of the paper gives basic information about all-things telephone, and some other handy information.

Our directories, because there is a significant amount of personnel turnover here (we're a k-8 school), "expire" twice a year. The expiration date is printed on the front of the phone directory. We expire the directory in March, and then again in August of the year. This expiration date prevents many versions of the directory to be floating about in the organization. It also keeps us on task to make certain that we do update it regularly.

What kind of information is presented on a phone directory? This depends entirely on the size of your organization. Our directory is divided by department. Smaller organizations may break their directories down by name, or position. For maximum usefulness, the layout and organization of telephone directories should reflect the scope and organization of the entity. Possible things to include on your directory are room numbers, job titles, and email addresses.

I have found that organizations that divide their names by department or job function find that the efficacy of the directory is improved. Part of the advantage of having a phone directory is for the initiation of the new staff, or volunteer base.

What about the back of the directory? Ours has notes on the back of the directory referring to several internal things, like how to file a work-order to have problems resolved.

There are basic instructions on changing personal settings, such as ring volume, tone, and personal options. Additionally, there are basic notes on how to use features of the telephone system, such as how to place a call on hold, and how to transfer a call. Voicemail, being a separate system, also has a set of instructions for basic usage.

Two notes which are of interest. We have a note on why not to move telephones (they won't work if moved). We also have a note about not trying to plug in "analog" telephone devices, such as modems or fax machines, into the jacks, because they can do damage to the system, and to the devices themselves.

* VISIT US ONLINE AT WWW.ST-JOSEPH-PARISH.ORG *

Figure 3-1: The Front and Back of our telephone directory

A BASIC TELECOM TOOL KIT

Like working on a car, a dishwasher, or a hair-dryer, there is very little which can be done without the aid of tools and test equipment.

Some tools in a basic telecom toolkit are obvious. You will need an assortment of slotted- and Phillips-type screwdrivers. The basic telecom toolkit here is designed for the casual system manager, and does not include tools necessary to pull wire in the walls, which is outside of the scope of this book. These tools are basically for making small repairs, such as replacing broken wall jacks, and testing parts of the wiring system. All of these items are available in electrical supply stores and consumer hardware stores.

A TELEPHONE TEST SET, A.K.A. "BUTT SET"

Figure 3-2: An ordinary telephone test-set

A "Butt-Set", or a Telephone test set is a fundamental tool of the telephone service trade, literally every lineman and phone company service rep will have one of these. Really, it is simply a heavy-duty telephone with a special pair of clips which allow the phone wires to clamp onto the small wires used to make communications circuits. This tool allows you to "butt in" on a regular phone line, thus the name. Not of much use at all to your digital phone system (it won't work), this tool allows you to hear what is going on in the phone-companies' wiring. These sets range in price, depending on manufacturer and features. As of this writing, and new, they start around $150, and average at around $300. It is very possible to find these items used online, and at swap meets for significantly less money.

PUNCHDOWN TOOL

This tool allows you to attach the wires to the patch panel, the block, or the jack installed in the wall. This tool is handy because it allows you to re-punch the wire down. More accurate and precise than a screwdriver, a punch-down tool does not damage wire or the jack. These tools range at around $45—100, depending on brand.

Figure 3-3: Example of a punch-down tool and included blades. Note the four ordinary blades. The top is for the "66" type blocks (left punches the wire down without cutting the wire, right side punches down the wire and cuts off the excess). The bottom blade is the 110 type, for wall jacks and 110-type blocks. As with the "66" blade, the left side punches down the wire, and does not cut off the excess, the right side punches down the wire and cuts off the excess. Note the visible cutting blade (pointed at top of the right hand side)

WIRE CUTTER/SCISSORS, AND CRIMP TOOL

Wire cutters and wire-cutting scissors are invaluable when making simple repairs to the wiring. These are commonplace, and inexpensive. A slightly different animal, the crimp tool, is a hand-tool which is used to put the ends on the wire. These tools are easy to use, and along with crimp ends, can be used to make cables such as those used to interconnect the equipment to the system. This tool can be used in a pinch to replace broken handset, station, and interconnection cable within your system.

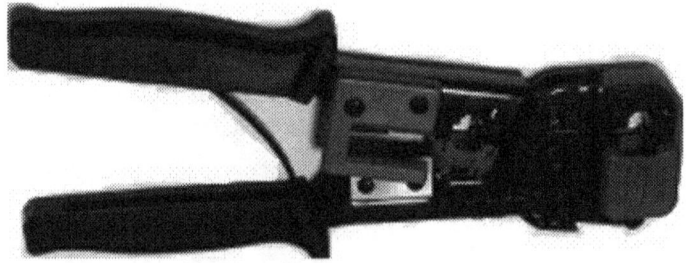

Figure 3-4:Crimp Tool

TONER/WAND (INDUCTIVE PROBE)

A Toner/Wand is a tool used to find a wire which has been installed, but left unlabeled, or a wire which is mislabeled. The Toner part generates an audible signal, the want is used on the other end to "find" the signal. This device set can be used to determine if there is a problem in the line itself. No-signal = no connection.

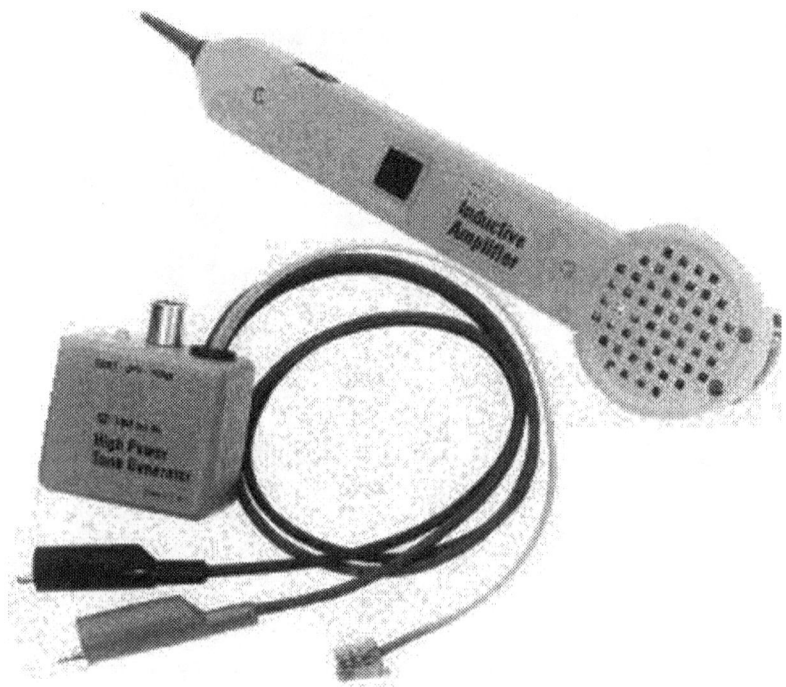

Figure 3-5: an example of a Toner/Wand (Inductive Probe) set

4-PAIR TESTER

A simple battery-operated test instrument, the 4-pair tester, can be used to test the connection between the wall jack and the panel in the equipment closet. This set has a "remote" (left), which generates a signal. The one on the right, the "Tester" is held by the individual doing the testing. This device is handy because it shows wiring faults, and mis-wired jacks which can go undetected with a toner/wand set, but will not appear to work.

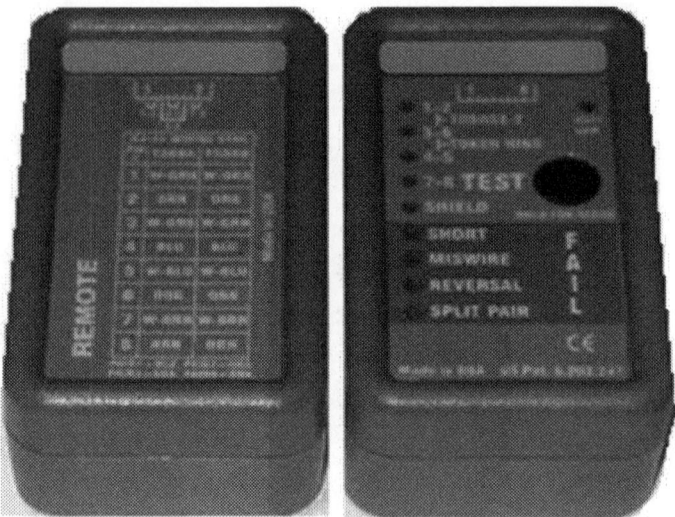

Figure 3-6: Example of a 4-pair tester and remote unit (left).

All of these simple tools are available through telecommunications and electronics retailers. As telecommunications and networking systems proliferate, these tools will likely become more readily available, and less expensive.

CONCLUSIONS, AND FAREWELL

Well, here we are on the other side of a lot of pages about telephones. I hope that you have learned a little about telecommunications, and not-for-profit organizations. Hopefully, it has given you a perspective on telecommunications system architecture, and some basics about how these systems work.

Moving forward? I highly recommend at least a cursory study of both the set of codes and requirements, and some other basic books about wiring before contracting with a vendor to do a wiring installation. Look also at product literature, which can be surprisingly informative about features, and the priorities that manufacturers place on features.

Take lots of notes, take care, and best wishes on the best of luck in your continuing journey!

APPENDIX

A SIMPLE NETWORK TOPOLOGY

Integration. This is a term that is vastly overused, and often misunderstood. Integration can mean a lot of things to a lot of different people. There is infrastructure integration, whereby the services in a building utilize the same "backbone" of wire. This kind of integration facilitates installation, and minimizes expense.

There is also feature integration, such as Integrated Voicemail. This integration implies that the equipment has been designed to interoperate at a software level for the benefit of the system managers or users of a system. This kind of automation also includes CTI, which involves the interconnection of computer systems and telephone equipment.

In this example, I am not speaking of integration on a software or feature level between computers and phones. That type of integration is available, but is highly dependent on the type of phone system you choose, and types of computers you install.

This integration type is at the wiring level. It's advantages are simple:

* Reduction in redundancy of hardware, and wire
* Increase in flexibility between computer and phone expansion
* Increase in ease and flexibility for moves, adds, and changes

Here we have a simple network model for integration of phone and data services can be implemented easily in your installation. The simplicity of this arrangement is that all house cabling wire is

(1) Punched down to the 568-B standard
(2) CAT5, 5e, or 6 (your preference, but not CAT3 telecom cable)
(3) punched down to one (or uniform) panel(s) using *one numbering schema.*

All components, like patch panels are CAT5, 5e, or 6 rated

All wall jacks are of a uniform color (e.g. black or beige), and are simply numbered using a uniform numbering scheme.

In some ways, this scheme goes against certain presumptions about structured cabling architecture. This method, though, allows a level of flexibility which is not matched by using separated, traditional, wiring schemes.

This scheme simplifies your network, because instead of having two discreet, and disconnected systems, your phone and data network share the house wiring and are integrated at the wiring level. Jacks in the offices, because they are all CAT5 or higher-rated cable, are suitable for either data or voice application.

Because the jacks are not different colors, numerous phones may be installed, or none at all, without disrupting the coding scheme. A simple list may be kept in the equipment closet, showing which ports are connected to which extensions, or which ports in the computer networking hub.

Additionally, utilizing a patch panel greatly simplifies making extension moves, because moving simply involves re-directing the wire coming from the phone system from the old to the new port on the patch panel.

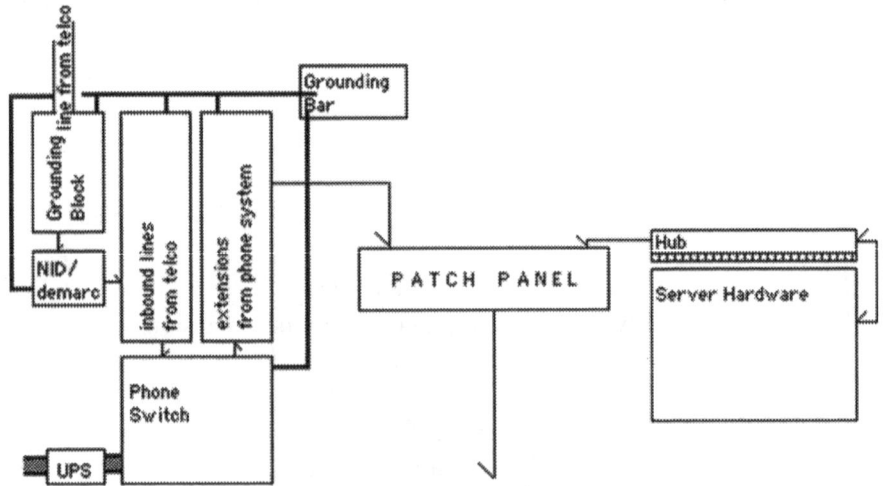

Figure A-1: Server hardware, and phone system hardware (shown in detail). The key to this schema is that phone and computer network wire share the house cabling system and same components. The phone system component connects to the network similar to the way that servers or hubs connect to the network.

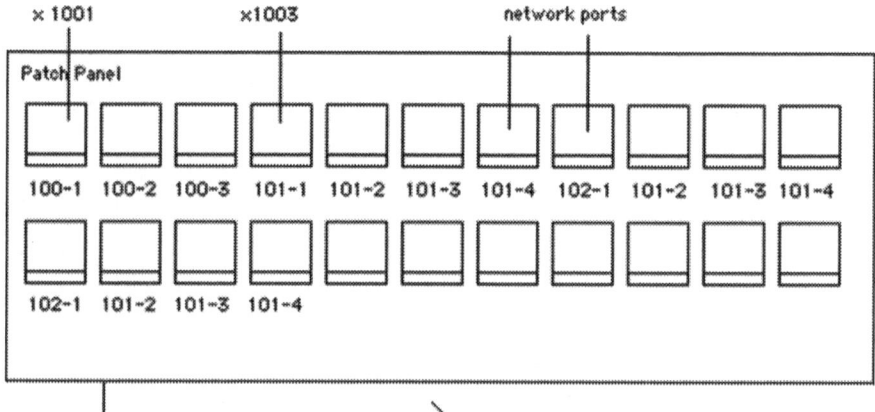

From back of the panel, to the house wiring and wall-jacks

Figure A-2: Example of a patch-panel. The uniform nature of the wiring scheme allows extensions and data ports to co-exist within one arrangement.

A listing of CHARTS and PICTURES

COVER/BACK/INTRODUCTION:

COVER	cover picture of telephones
BACK	back cover photo
INSIDE	Inside Cover picture

SECTION 1:

1-1	Example of a Key-type desk telephone
1-2	Anatomy of a typical PBX telephone
1-3	Example of a PBX desk telephone
1-4	Example of a medium-sized phone switch
1-5	Example of small-scale phone switch
1-6	A basic wiring diagram
1-7	A Simple Cross-connect diagram
1-8	EIA/TIA standard equipment room
1-9	Phone System Management Chart
1-10	Backboard diagram
1-11	66- and 110-type blocks
1-12	Graph of ACR
1-13	568-A and B wire mapping schema
1-14	Picture of jack label

SECTION 2:

SECTION 3:

APPENDIX 1:

0-595-32385-5